Mathematical Principles for Scientific Computing and Visualization

Mathematical Principles for Scientific Computing and Visualization

Gerald Farin
Dianne Hansford

A K Peters, Ltd.
Wellesley, Massachusetts

Editorial, Sales, and Customer Service Office

A K Peters, Ltd.
888 Worcester Street, Suite 230
Wellesley, MA 02482
www.akpeters.com

Library of Congress Cataloging-in-Publication Data

Farin, Gerald E.
Mathematical principles for scientific computing / Gerald Farin, Dianne Hansford
p. cm.
Includes bibliographical references and index.
ISBN 13: 978-1-56881-321-9 (alk. paper)
1. Science–Data processing. 2. Numerical analysis–Data processing. 3. Information visualization–Data processing. 4. Computer graphics. I. Hansford, Dianne. II. Title.
Q183.9.F37 2008
502.85–dc22

2008002537

Cover image courtesy of Scientific Computing and Imaging Institute, University of Utah; see Figure 15.17.

Printed in India
12 11 10 09 08 10 9 8 7 6 5 4 3 2 1

Contents

Preface xi

1 Introduction 1

2 Computational Basics 5
- 2.1 Algorithms 5
- 2.2 Floating-Point Numbers 6
- 2.3 Errors 7
- 2.4 Case Study: The 1991 Scud Attack 10
- 2.5 Problems and Experiments 11

3 Coordinate Systems 13
- 3.1 Cartesian Coordinate Systems 13
- 3.2 Polar, Spherical, and Cylindrical Coordinate Systems 16
- 3.3 Case Study: UTM Coordinates 19
- 3.4 Local and Global Coordinates 21
- 3.5 Homogeneous Coordinates 25
- 3.6 Problems and Experiments 26

4 Background: Numerical Linear Algebra 27
- 4.1 Linear Spaces and Vectors 27
- 4.2 Linear Independence and Subspaces 29

4.3 Linear Maps and Matrices 31
4.4 Lengths and Volumes 37
4.5 Summary of Matrix Rules 40
4.6 Problems and Experiments 40

5 Solving Linear Systems 41
5.1 Case Study: Mixing Chemicals 41
5.2 Linear System Basics 42
5.3 Gauss Elimination 43
5.4 Stability 45
5.5 Vector Norms and Sequences 46
5.6 Iterative System Solvers 48
5.7 Case Study: Fluid Flow 51
5.8 Overdetermined Systems 53
5.9 Case Study: Femoral Head Reconstruction 54
5.10 Problems and Experiments 55

6 Eigen-Problems 57
6.1 Eigenvalues 58
6.2 The Power Method 60
6.3 Case Study: PageRank 61
6.4 Jacobi Iteration 63
6.5 Eigenvalues and Determinants 66
6.6 Singular Value Decomposition 67
6.7 The Condition Number 70
6.8 The Pseudoinverse 70
6.9 The Principal Components Analysis 72
6.10 Case Study: Eigenfaces 73
6.11 Singular Values, Volumes, and Determinants . . . 75
6.12 Problems and Experiments 76

7 Background: Numerical Calculus 77
7.1 Functions 77
7.2 Limits 80
7.3 Integrals 81
7.4 Derivatives 83
7.5 Function Spaces 86
7.6 Problems and Experiments 88

8 Data Fitting 91
8.1 Taylor Approximation 91
8.2 Piecewise Linear Interpolation 92
8.3 Polynomial Interpolation 92
8.4 Polynomial Least Squares Approximation 94
8.5 General Least Squares Approximation 98
8.6 B-Spline Interpolation 99
8.7 B-Spline Least Squares Approximation 104
8.8 Integrals and Derivatives 105
8.9 Problems and Experiments 107

9 Computing Dynamic Processes 109
9.1 Background 109
9.2 Euler's Method 111
9.3 Heun's Method 113
9.4 Boundary Values 115
9.5 ODEs and Dynamical Systems 117
9.6 Case Study: The Lorenz Attractor 120
9.7 Problems and Experiments 122

10 Finding Roots 123
10.1 The Piecewise Linear Approach 123
10.2 The Newton-Raphson Method 125
10.3 Case Study: Computing the Square Root 127
10.4 Bisection 128
10.5 Case Study: Wilkinson Polynomials 129
10.6 Problems and Experiments 129

11 Computing with Multivariate Functions 131
11.1 Bivariate Functions 131
11.2 Bilinear Interpolation 135
11.3 Quadratic Forms 137
11.4 Contouring 138
11.5 The Newton-Raphson Method 140
11.6 Partial Differential Equations 142
11.7 Trivariate Functions 145
11.8 Problems and Experiments 147

12 Visualizing Empirical Data 149
12.1 Scatter Plots, Correlations, and Regression . . . 149
12.2 PCA Revisited . . . 153
12.3 Histograms, Bar Charts, and Pie Charts . . 155
12.4 Box Plots . . . 159
12.5 Log Plots . . . 161
12.6 Problems and Experiments . . . 163

13 Facets 165
13.1 Triangles . . . 165
13.2 Barycentric Coordinates . . . 167
13.3 Planes . . . 169
13.4 Polygons and Polyhedra . . . 172
13.5 Triangle Meshes . . . 174
13.6 Case Study: 3D Archiving . . . 178
13.7 Analyzing Triangle Meshes . . . 179
13.8 Delaunay Meshes and Voronoi Diagrams . . . 181
13.9 Case Study: Bark Beetles . . . 184
13.10 Other Meshes . . . 184
13.11 Problems and Experiments . . . 186

14 Visualizing Scalar Values over 2D Data 189
14.1 Height Maps . . . 192
14.2 Color Maps . . . 194
14.3 Contours . . . 197
14.4 Case Study: GIS . . . 204
14.5 Image Segmentation . . . 206
14.6 Problems and Experiments . . . 208

15 Volume Visualization 209
15.1 Scalar Data over a Volume . . . 209
15.2 Contouring . . . 215
15.3 Case Study: Health Care . . . 218
15.4 Direct Volume Rendering . . . 219
15.5 Transfer Functions . . . 222
15.6 Comparison of Contouring and DVR . . . 224
15.7 Case Study: Visible Human Project . . . 225
15.8 Data Cutting . . . 226

15.9 Vector Fields 227
15.10 Tensor Fields 229
15.11 Haptic Visualization 230
15.12 Problems and Experiments 231

16 Background: Computer Graphics 233
16.1 Color Models 234
16.2 The Graphics Pipeline 236
16.3 Illumination Models 252
16.4 Texture Mapping 261
16.5 Sampling, Smoothing, Reduction 263
16.6 Problems and Experiments 266

Bibliography 269

Index 271

Preface

Given the current explosion of data gathering and storage, we face a shortage of trained scientists and engineers who are able to extract knowledge from such data. In order to solve the big problems of today and tomorrow, we need more people with adequate training, and we need to create better tools for extracting knowledge from huge collections of data. Building a new generation of tools will require interdisciplinary teams that can think in new ways. In order for these teams to be productive, they require a common language, and the fundamentals of scientific computing and visualization (SCV) form this language. Thus, the primary goal of *Mathematical Principles for Scientific Computing and Visualization* is to bring SCV tools to a diverse group of people.

Recently, a new trend has started in computer science departments: the move to bring computer- and technology-based knowledge to a broader group of people. This movement is called *informatics.*[1] Computer applications are ubiquitous today, influencing nearly every aspect of our lives. Computer scientists and engineers alone cannot keep pace with the developments. Thus, informatics departments are emerging at colleges and universities to train a new breed of specialists.

[1]This is a US term. In Europe, "informatics" is synonymous with "computer science."

Mathematical Principles for Scientific Computing and Visualization was written with the goals of informatics in mind. In our opinion, it is the first book on SCV that targets such a broad audience. It is intended for students and researchers in engineering and science-related areas, such as biology, geography, and psychology. It will appeal to students because it concisely covers a range of important topics from an application-oriented perspective. Professionals will find this book helpful as a review of key computational methods, or as an update to what they were taught. Individuals in either of these audiences, in the course of their professional applications or research pursuits, will be exposed to various software packages for solving problems, be it problems from statistics, applied mathematics, or scientific visualization, in addition to domain-specific software. This book is intended as a guide to understanding the mathematical principles that underly the more general software packages. That knowledge is important for the non-mathematician because naive and uneducated use of computing and visualization packages might produce meaningless or erroneous results.

This book is written in an informal style and is accessible to someone who is not a mathematician. The book has over 180 illustrations —and not only in the "visualization" part. The reader is led to understand many concepts through graphical examples. Practical suggestions for using the tools of SCV are given, and applications are described in the text and demonstrated with illustrations. Case studies, real-world examples of how one or more tools are used, are included for nearly all topics. Positive feedback on other book projects convinced us that this style of book is of great use to practitioners and people new to a field. (See http://www.farinhansford.com/books.html for a complete list of books by the authors.)

Review of Contents

The book has two basic parts: scientific computing (Chapters 4–11) and visualization (Chapters 12–16). The book is designed so that the reader can begin with either part; however, cross references are given when material is dependent on ideas discussed elsewhere. The first two chapters after the introduction, "Computational Basics" and "Coordinate Systems," introduce the reader to key concepts that are used throughout the book.

The part of the book that focuses on scientific computing begins with topics on numerical linear algebra with Chapters 4, 5, and 6: "Background: Numerical Linear Algebra," "Solving Linear Systems," and "Eigen-Problems."

The second component of the scientific computing part, Chapters 7–11, deals with numerical calculus topics: "Background: Numerical Calculus," "Data Fitting," "Computing Dynamic Processes," "Finding Roots," and "Computing with Multivariate Functions."

The visualization part of the book begins with the most basic tools, which are presented in Chapter 12: "Visualizing Empirical Data." In order to develop more advanced visualization tools, "Facets" is the next chapter, and it focuses on triangle meshes. Chapters 14 and 15, "Visualizing Scalar Values over 2D Data" and "Volume Visualization," present state-of-the-art visualization techniques. For deeper knowledge of visualization, the last chapter, "Background: Computer Graphics," provides details on how objects are rendered in the visualization process.

Each chapter concludes with a "Problems and Experiments" section. The problems are not rote calculations, but rather require some reflection on the topics presented. Experiments are designed to incorporate the use of a software package and thus give the reader hands-on experience with the methods. Only through experimentation does one realize the power and pitfalls of systems such as Mathematica, Matlab, and Maple.

Classroom Use

In a classroom setting, *Mathematical Principles for Scientific Computing and Visualization* targets junior or senior undergraduates. A background in basic computing skills is desirable, as well as some basic knowledge of calculus and linear algebra.

For a one-semester class (and for an audience of varying mathematics backgrounds), the first chapters on computing and coordinates, Chapters 2 and 3, are essential for forming the necessary foundation. If the students have a good mathematical background, then all chapters from Chapter 4 (linear algebra) to Chapter 11 (multivariate data) can be treated in depth, and those on visualization may be given a lighter treatment. Conversely, if students are computationally oriented, there ought to be an emphasis on the visualization part, Chapters 12–16, and a lighter treatment of the scientific computing part.

Website

The book's website is http://www.farinhansford.com/books/scv/index.html. The website contains teaching materials, as well as the figures and code used in the book. We used Mathematica for computations and for generating many of the figures in the book. Yet this is not a Mathematica-centered book: the text is designed so that readers may equally well use other packages such as Matlab or Maple. Reviews and errata will be posted on the site as well.

Acknowledgments

We want to thank the many people and organizations that gave permission for us to use their images. We credit the many contributors in source lines that appear at the ends of figure captions.

Again, thanks to the great team at A K Peters! They are a pleasure to work with.

Gerald Farin
Dianne Hansford
Arizona State University
May 2008

1

Introduction

The goal of science is the creation of knowledge. The process typically starts with raw data, which are processed to extract information. Interpretation of this information then leads to knowledge. Roughly speaking, scientific computing aids in the first step, whereas scientific visualization helps in the second one. Figure 1.1 illustrates these steps.

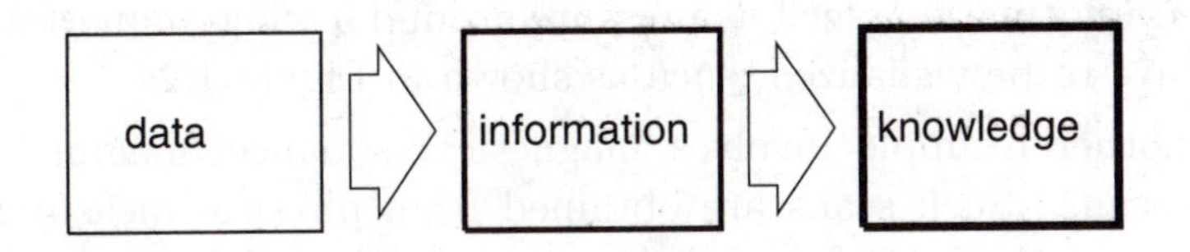

Figure 1.1. Steps in the scientific process.

Scientific computing and visualization are important elements of the iterative process of scientific discovery. Both tools provide the researcher with more information for refining a hypothesis, building a better mathematical model for abstracting a phenomenon, testing data acquisition methods, and evaluating observations. Visualization, through transformation and rendering, provides additional validation and verification of each process.

As an example, imagine the process for developing a new airplane by an engineering team at a major aircraft company. Before a prototype is built, extensive computer simulations take place, typically

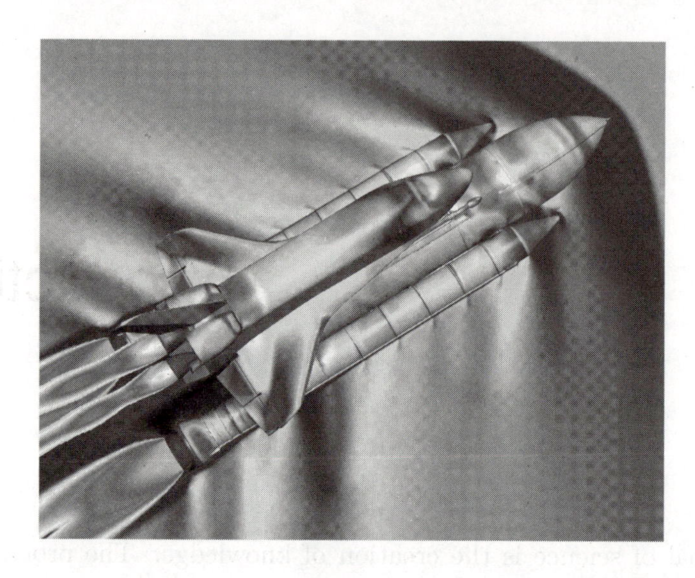

Figure 1.2. Simulation of air flow around the space shuttle. (Image courtesy of NASA.)

involving numerical solutions of partial differential equations. These simulations yield large amounts of numerical data. To extract the desired information, such as pressure around a wing, numerical data sets have to be visualized, such as shown in Figure 1.2.

Another example involves magnetic resonance imaging (MRI) brain scans. Such scans are obtained from physical measurements (data) and the use of complex numerical algorithms (creating information). To interpret these scans, they must be visualized by transforming them into images such as the one shown in Figure 1.3 (creating knowledge).

Therefore, visualization may be thought of as the graphical depiction of data and information. Instead of being one monolithic discipline, visualization takes on several forms:

- *Scientific visualization* focuses on scientific data and mathematical modeling techniques. Most often, this discipline represents spatial or natural geometric information with physical attributes attached.

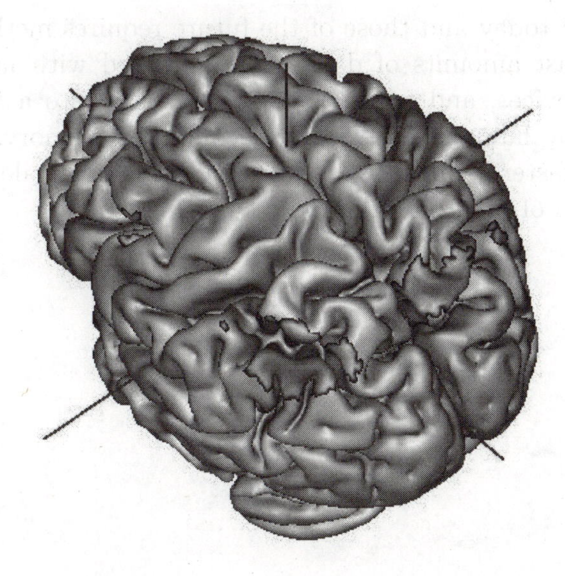

Figure 1.3. An MRI image of a human brain. (Image courtesy of Arizona Alzheimer's Disease Consortium.)

- *Information visualization* mostly focuses on nonspatial, abstract data. This discipline receives credit for many human-computer interface, cognitive, and perception advances in the area of graphical representations.
- *Visual analytics* focuses on analytical reasoning supported by visualization. This is a new area, formed to bring together tools from scientific visualization and information visualization.

With each of these disciplines, the goal is to leverage the human brain's strong dependency on sight to allow for a better understanding of a given problem or phenomenon. (In sighted individuals, nearly one-third of the brain is devoted to processing visual information.) The focus of this book is scientific visualization, recognizing that the boundaries between the disciplines are not clearly defined.

Visualization is an interdisciplinary field, bringing together many domain sciences: scientific computing, computer graphics, image processing, human-computer interfaces, and cognitive science.

Scientific computing and visualization are effective tools for building links among data, information, and knowledge. Solving the big problems of today and those of the future requires methods to deal with the vast amounts of data being gathered with new data acquisition devices, and transforming these data into a form that is digestible by humans. Inexpensive computer memory, specialized chips, and faster computers are allowing for larger-scale simulations in the realm of high-performance computing.

2

Computational Basics

Scientific computing (also referred to as computational science) solves scientific problems by constructing good mathematical models for these problems and then using efficient computational tools to find solutions. Within the field of computer science, these algorithms are characterized by the fact that they are almost entirely based on computing with floating point numbers. Here, we give a brief overview of the concepts of algorithms and floating-point numbers.

2.1 Algorithms

Scientific computing is about executing *algorithms*, typically encountered in the form of computer programs. An algorithm is a step-by-step set of instructions for solving a particular problem, much like a recipe tells how to prepare a meal.

An algorithm must be *unambiguous.* For example, "Execute either instruction X or instruction Y" does not clearly specify which instruction to execute. However, "If $a > 0$, execute X, and execute Y otherwise" is unambiguous.

An algorithm must *terminate.* This means no infinite loops are allowed. In reality, it is almost impossible to check ahead of time whether a program (the computer implementation of an algorithm) can avoid infinite loops.

An algorithm must be *complete*. All possible circumstances must be covered. In practice, one overlooked circumstance can prove deadly!

An important attribute of algorithms is their *complexity*. Some algorithms may just be badly designed, making their execution time horrendously long even if the problem to be solved is not very complex. However, there are problems for which efficient algorithms do not exist. Those problems are called *hard*. For instance, solving n linear equations with n unknowns roughly requires n^3 computations—this is considered not hard. But if the equations become nonlinear, no bound on the number of computations can be given—hence solving systems of nonlinear equations is very hard.

2.2 Floating-Point Numbers

The TV news might report that tomorrow's temperature is expected to be 48 degrees. The news might also report that the Dow Jones index is at 12,335.47 points. These two pieces of information come at different levels of detail: one (temperature) gives two digits; the other (stock prices) gives seven digits. It appears that the Dow Jones number is inflated with superfluous information (or, taking an opposing point of view, the temperature information is lacking detail). At a two-digit level of information, we might say the Dow Jones index is at 12,000 points—with this approximation, we miss some detail but we do get the overall picture. Conversely, we would not think of a seven-digit forecast of 48.10482 degrees as meaningful—this would be information overload.

The concept of keeping the essential digits and discarding "noise" is key to understanding *floating-point numbers*. You may encounter them in various forms, such as

$$3.14159, \quad 0.314159 \text{ E } 01, \quad .314159 \text{ E } 01, \quad +.314159 \text{ E } 01, \quad 31.4159 \text{ E } -01,$$

which all denote the same number. The first one is standard fare; the remaining ones are variations on a common form $xxx.yyy$ E $\pm zz$. Here again the $xxx.yyy$ part is a standard number; the E $\pm zz$

part indicates that $xxx.yyy$ is multiplied by $10^{\pm zz}$.[1] This notation is particularly useful when dealing with very large or very small numbers. The following two notations are equivalent:

$$0.000000001, \quad 0.1 \text{ E } -08,$$

and so are these:

$$7000000000, \quad 0.7 \text{ E } 10.$$

The exponent notation is much easier to comprehend!

There are infinitely many real numbers, but a computer can deal only with finitely many. Thus, floating-point numbers only approximate the much larger set of reals. Floating-point numbers are typically represented as

$$0.x_1 x_2 x_3 \cdots x_{16} \text{ E } \pm e_1 e_2,$$

following the IEEE[2] standard for 64-bit number representation. This means a floating-point number is stored as 16 digits $x_1 \cdots x_{16}$, followed by a 2-digit exponent. The range for the exponent is about -38 to 38. Numbers exceeding 10^{38} in absolute value are treated as NaN (not-a-number).

2.3 Errors

When computing with floating-point numbers using digital computers, we cannot always expect 100% correct results. Mostly, we do not care and are satisfied with results that are accurate to 10 digits or so. But even those 10 digits cannot always be guaranteed to be accurate! The different kinds of numerical errors and their effect on computations are the topics of numerical analysis. We shall give some examples.

Example 1. The quadratic equation

$$x^2 - 2 = 0$$

[1] The "E" symbol stands for "exponent."

[2] Institute of Electrical and Electronics Engineers; this organization is responsible for creating many computer-related international standards.

is known to have two exact solutions,

$$x_{1,2} = \pm\sqrt{2}.$$

However, there is no way a digital computer can *exactly* compute $\sqrt{2}$ in floating-point form. It may be computed to high accuracy (typically to 16 digits), but the exact value requires infinitely many digits in its decimal expansion—a computer cannot handle that. Hence, square roots have to be approximated within some tolerance, giving rise to numerical errors.

Example 2. The sine function is defined as

$$\sin(x) = \sum_{i=1}^{\infty} \frac{(-1)^{i-1}}{(2i-1)!} x^{2i-1}.$$

For practical computations, this expression cannot be evaluated for infinitely many terms. We have to terminate (truncate) for some finite value n (instead of ∞), resulting in a numerical error that may be considerable for large x; the truncated sum is a polynomial of degree n in x. The sine function is bounded, whereas no polynomial is! Errors arising this way are referred to as *truncation errors.*

Example 3. The number $1/10$ is well-defined in a decimal context. A digital computer, however, needs to convert it by using powers of 2:

$$\frac{1}{10} = .0001100110011...$$

Realistically, this infinite expansion has to be truncated somewhere, thus $1/10$ cannot be accurately represented by a digital computer. This leads to the surprising fact that 10×0.1 will not result in the exact value 1.0; instead it will be missed by about 10 E -16. See Section 2.4 for more on this.

Example 4. The function

$$f(x) = \frac{1 - \cos x}{x^2}$$

is plotted in Figure 2.1.

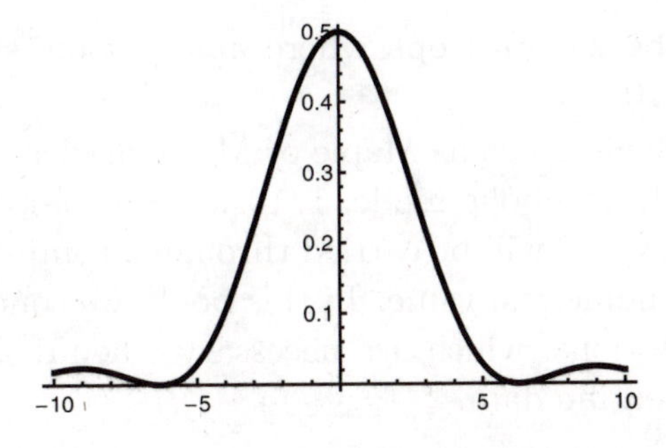

Figure 2.1. The function $(1 - \cos x)/x^2$ over the interval $[-10, 10]$.

It can be shown that $f(0) = 0.5$; in the close vicinity of $x = 0$, the function behaves like the constant $y = 0.5$. Yet when we plot f over a very small interval $[-5.0 \text{ E } -8, 5.0 \text{ E } -8]$ (Figure 2.2), the plot looks fairly disturbing. The reason for the difference is the use of floating-point numbers: near $x = 0$, the terms 1 and $\cos x$ are almost identical. Their difference will yield only very few meaningful digits, resulting in computations that use only two or three relevant digits. This kind of error is referred to as *cancellation error.* It is responsible for the erratic behavior of the second plot.

Example 5. Because of the possibility of numerical errors, *never* check for equality of floating-point numbers! Even if equality is expected, it typically is not met exactly. Thus, instead of checking

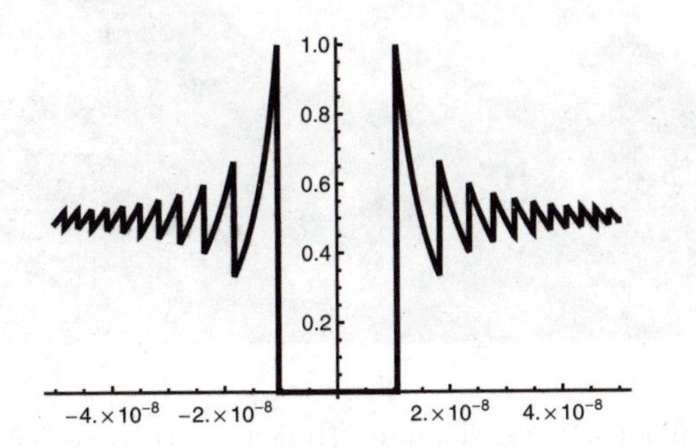

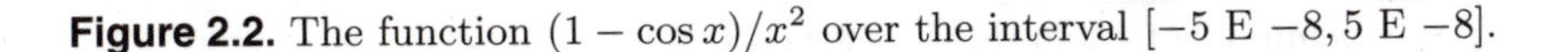
Figure 2.2. The function $(1 - \cos x)/x^2$ over the interval $[-5 \text{ E } -8, 5 \text{ E } -8]$.

`x==y`,[3] check `abs(x − y) < eps` where `eps` is some small tolerance, such as 10 E −10.

Several packages, such as Maple or Mathematica, allow symbolic computations. In symbolic mode, $1/10$ will not be converted to approximate binary; $\sqrt{2}$ will be carried through a computation without conversion to a numerical value. In this book, we concentrate on numerical computations, which are necessary when dealing with data arising from scientific data.

2.4 Case Study: The 1991 Scud Attack

The 1991 Scud attack is a case in which human lives were lost due to numerical error.

On February 25, 1991, during the first Iraq war, 28 American soldiers were killed by an Iraqi Scud missile because the "Patriot" defense system failed to intercept it—see Figure 2.3 for a view of a Patriot unit.

The reason for this death toll was numerical error. The Patriot unit kept its own internal clock, updating it every tenth of a second starting from reboot. "One-tenth" was stored as a 24-bit binary number which misses the true value 0.1 by about 0.000000095.

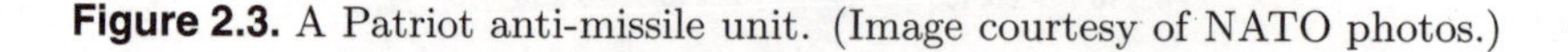

Figure 2.3. A Patriot anti-missile unit. (Image courtesy of NATO photos.)

[3] The `==` symbol is frequently used for an equality query.

An incoming missile is first detected by a radar system—and this system worked using the correct time, stored as a floating-point number. If the detected object were indeed a missile, following the laws of ballistic flight, then its path would be predictable. In particular, a fraction of a second after initial detection, it must appear at a location that is quickly computed by using elementary physics. The Patriot software checks whether the detected object does in fact appear at the predicted location at the predicted time. If it does not, it is interpreted as not being a ballistic missile and no action is taken. The reason for this safeguard is that a Patriot missile had a price tag of about $600,000, so firing one should not happen without a confirmed threat at hand.

At the time of the Scud attack, the Patriot system was up for about 100 hours, and due to the accumulated time error, it was 0.34 seconds off actual time. A Scud travels about 600 meters during that time span. So when the Patriot software checked whether the detected object was in fact a ballistic missile, it miscalculated the predicted location by 600 meters and detected no object at that location. A nonballistic object was assumed, no Patriot was launched, and 28 people died.

2.5 Problems and Experiments

1. Define two floating-point numbers by

$$x = 1.0 \text{ E } 20 \quad \text{and} \quad y = 1.0 \text{ E } -20.$$

   Neither x nor y equals zero, and thus we would expect $x - y \neq x + y$. However, when executed in software, you will most likely obtain equality. Experiment yourself! Explain why.

2. Find three floating-point numbers x, y, z such that

$$x + (y + z) \neq (x + y) + z.$$

   Hint: these numbers should vary considerably in magnitude.

3. Using the IEEE conventions outlined in Section 2.2, how many floating-point numbers are there?

4. How many floating-point numbers are in the interval $[-10^{-10}, 10^{-10}]$ and how many are in the interval $[10^{10} - 10^{-10}, 10^{10} + 10^{-10}]$?

5. Experiment with the two functions

$$f(x) = \sin\frac{1}{x} \quad \text{and} \quad g(x) = x\sin\frac{1}{x}.$$

They behave very differently near $x = 0$. Plot the functions and describe their behavior.

3

Coordinate Systems

When dealing with objects in space in a computational setting, we need to know where the objects are. This means that, relative to a known reference system, they are identifiable by *coordinates.* Many types of coordinate systems exist; which one to use depends on the particular application. In some cases, we need to transition from one coordinate system to another, resulting in *coordinate transformations.*

3.1 Cartesian Coordinate Systems

The *Cartesian coordinate system* forms the foundation for all the topics of this book. It has been said that René Descartes (1596–1650), who is credited with inventing this system, did so as he lay ill in bed. Looking up, he noticed a fly moving around on the ceiling which was made of tiles. Descartes realized that he could describe the position of the fly with respect to the tiling. Thus was born the Cartesian coordinate system.

Illustrated in Figure 3.1 is a two-dimensional (2D) Cartesian coordinate system, which provides a reference frame by defining an *origin* and perpendicular coordinate axes called the x- and y-axes. The axes are marked with chosen units of length, which may or may not be the same for both axes. These elements together are called a *coordinate plane.* Nearly everyone follows the *right-hand rule* convention for axes orientation. First, establish the positive x-axis. Next,

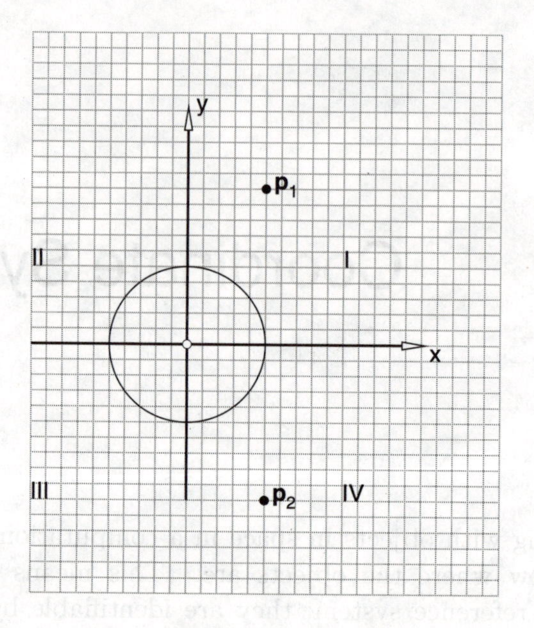

Figure 3.1. The 2D Cartesian coordinate system. Point $\mathbf{p}_1$ has (x, y)-coordinates $(1, 2)$. Point $\mathbf{p}_2$ has (x, y)-coordinates $(1, -2)$. The four quadrants are labeled I–IV.

align your right hand so your fingers point in the direction of the $+x$-axis with your thumb pointing up. Then the $+y$-axis will be in the direction that your fingers curl.

Coordinate systems are important tools for mathematics because they allow a unique determination of a location (point) by its *coordinates.* Three points are illustrated in Figure 3.1: the origin, $\mathbf{p}_1$, and $\mathbf{p}_2$. Furthermore, coordinate systems are important because they lend themselves to algebraic equations for geometric entities such as the circle $x^2 + y^2 = 1$, which is also illustrated in the figure.

A convenient convention for referring to groups of geometric entities or measurements with the same pair of x- and y-coordinate signs is to use the *quadrant* notation. Labeled in Figure 3.1 by Roman numerals I–IV are the four quadrants of the 2D system.

In this book, as we develop more tools for representing geometry and scientific measurements, we'll find it convenient to have a more general notation for coordinate systems, so let's refer to the origin as $\mathbf{o}$ and the axes as $\mathbf{e}_1$ and $\mathbf{e}_2$. Using this notation, we to define the

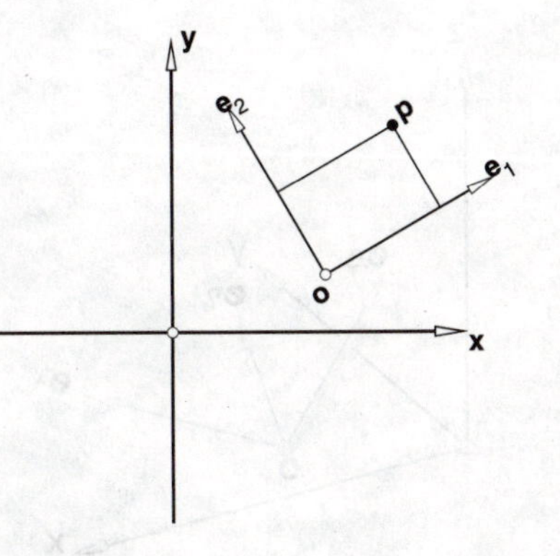

Figure 3.2. A more flexible notation for the 2D Cartesian coordinate system. This allows us to define geometry with respect to the frame anywhere in the plane.

2D Cartesian coordinate system as

$$\mathbf{o} = \begin{bmatrix} 0 \\ 0 \end{bmatrix}, \quad \mathbf{e}_1 = \begin{bmatrix} 1 \\ 0 \end{bmatrix}, \quad \mathbf{e}_2 = \begin{bmatrix} 0 \\ 1 \end{bmatrix}. \tag{3.1}$$

A position $\mathbf{p}$ in the plane is defined by its coordinates (p_1, p_2) in this frame, thus

$$\mathbf{p} = \mathbf{o} + p_1\mathbf{e}_1 + p_2\mathbf{e}_2.$$

We use this more flexible $\mathbf{o}$, $\mathbf{e}_i$ notation so we can place our reference frame anywhere, as illustrated in Figure 3.2. Notice that we have expressed the coordinate frame and position $\mathbf{p}$ in the figure as a column vector; more information on this is given in Section 4.1. Exactly how to construct the transformation in the figure is explained in Section 4.3.

The three-dimensional (3D) Cartesian coordinate system is created from the 2D system by adding a z-axis. Thus a point is defined by three coordinates, (x, y, z). As with the 2D system, we will follow the right-hand rule for orientation of the axes. The direction of your thumb indicates the positive z-axis direction, as Figure 3.3 illustrates. Again, to allow for more flexibility, we use the $\mathbf{o}$, $\mathbf{e}_i$

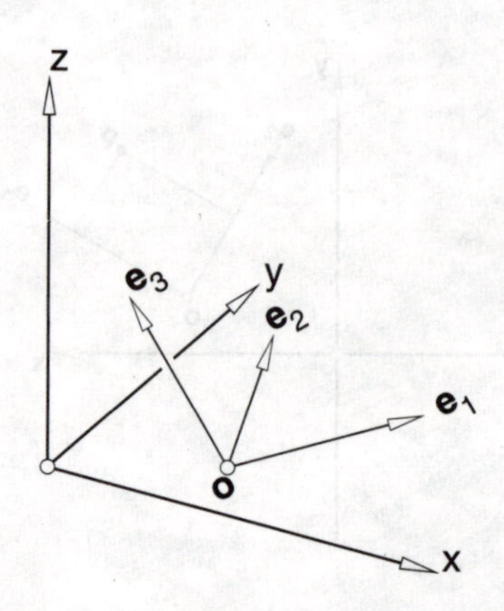

Figure 3.3. The 3D Cartesian coordinate system.

notation. To represent the Cartesian coordinate frame, we let

$$\mathbf{o} = \begin{bmatrix} 0 \\ 0 \\ 0 \end{bmatrix}, \quad \mathbf{e}_1 = \begin{bmatrix} 1 \\ 0 \\ 0 \end{bmatrix}, \quad \mathbf{e}_2 = \begin{bmatrix} 0 \\ 1 \\ 0 \end{bmatrix}, \quad \mathbf{e}_3 = \begin{bmatrix} 0 \\ 0 \\ 1 \end{bmatrix}. \tag{3.2}$$

A position $\mathbf{p}$ in the plane is defined by its coordinates (p_1, p_2, p_3) in this frame; thus,

$$\mathbf{p} = \mathbf{o} + p_1\mathbf{e}_1 + p_2\mathbf{e}_2 + p_3\mathbf{e}_3.$$

3.2 Polar, Spherical, and Cylindrical Coordinate Systems

Let's look at some other commonly used coordinate systems and examine why we need them.

The *polar coordinate system* is a 2D coordinate system that defines a point with two coordinates (r, θ), where r is a *radial coordinate* and θ is an *angular or azimuthal coordinate.* As illustrated in Figure 3.4, it is standard practice to align the polar coordinate system with the Cartesian frame by associating the $+x$-axis with $0°$. As we travel counterclockwise, through quadrants I–IV, the angular coordinate increases to $360°$. The angular coordinate can be expressed in degrees or radians.

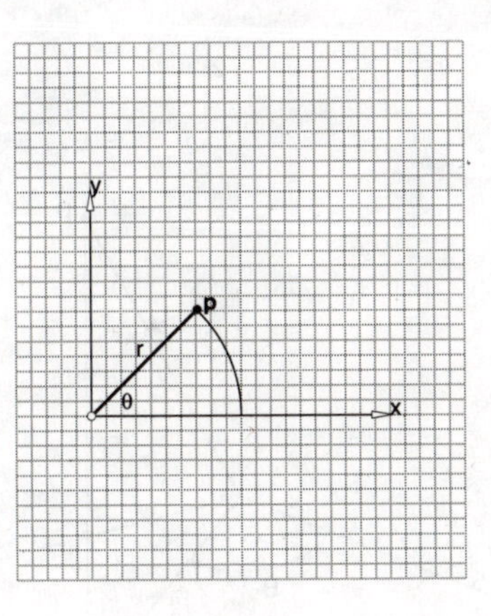

Figure 3.4. The polar coordinate system. The point $\mathbf{p}$ has polar coordinates $(2, 45°)$.

Negative angular coordinates are formed by traveling clockwise about the origin. Normally the radial coordinate is positive; however, it is possible to define a negative radial coordinate; it is interpreted as representing a point in the quadrant opposite that of the point with the same positive radial coordinate. To avoid redundancy, a good practice is to constrain the coordinates as follows:

$$r \geq 0 \quad \text{and} \quad 0° \leq \theta < 360°.$$

However, if important information is contained in the number of rotations, then this constraint on θ should not be implemented.

One way in which polar coordinates differ from Cartesian coordinates is that one point can have many coordinates. For example, the following polar coordinates represent the same point: $(2, 45°)$, $(2, 405°)$, $(2, -315°)$.

Obviously, polar coordinates are suited to modeling phenomena having a rotational element or a center point with a radial definition. Cam profiles are a good example. Also, gravitational or flow problems can be simpler to formulate in polar coordinates.

The 3D version of polar coordinates is called *spherical coordinates.* These coordinates allow us to identify a point in 3D with three coordinates (r, θ, ϕ), where r is a radial coordinate, θ is an azimuthal (angular) coordinate, and ϕ is a zenith (angular) coordinate. As illustrated in Figure 3.5, spherical coordinates identify a

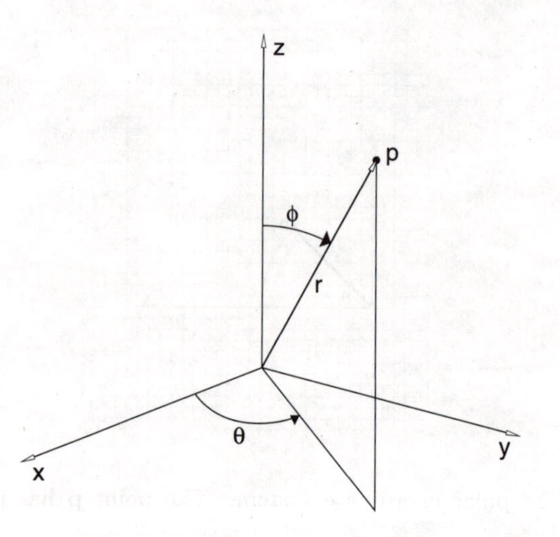

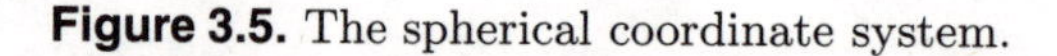

Figure 3.5. The spherical coordinate system.

point on the equator of a sphere with radius r, and ϕ identifies a latitude (north-south) measured from the $+z$-axis. The domains for the coordinates are defined as

$$r \leq 0, \qquad 0^\circ \leq \theta < 360^\circ, \qquad 0^\circ \leq \phi \leq 180^\circ.$$

To remain on one sphere, say Earth, we hold r constant and let the azimuth and zenith vary. Spherical coordinates are then easily converted to the familiar *latitude and longitude.* Recall that latitude, commonly abbreviated as Lat, has a domain of $+90^\circ$ at the north pole, 0° at the equator, and -90° at the south pole. Points on the sphere with constant latitude are called parallels. Longitude, commonly abbreviated as Long, has a domain of 0° at Greenwich (near London, UK); traveling east, the coordinate increases to $+180^\circ$; traveling west, the coordinate decreases to -180°. Points on the sphere with constant longitude are called meridians. The meridian of $\pm 180^\circ$ longitude is called the prime meridian (and it passes through the Fiji Islands). If we assume that the $+x$-axis runs through the Greenwich meridian, then the conversion from spherical coordinates to Lat and Long is as follows:

$$\text{Lat} = 90^\circ - \phi \quad \text{and} \quad \text{Long} = \begin{cases} \theta & \text{if } \theta \in [0^\circ, 180^\circ], \\ \theta - 360^\circ & \text{if } \theta \in (180^\circ, 360^\circ]. \end{cases}$$

Conversions from spherical coordinates to Cartesian coordinates are

$$x = r\cos\theta\sin\phi, \qquad y = r\sin\theta\sin\phi, \qquad z = rcos\phi.$$

Conversions from Cartesian coordinates to spherical coordinates are

$$r = \sqrt{x^2+y^2+z^2}, \qquad \theta = \arctan y/x, \qquad \phi = \arccos z/\sqrt{x^2+y^2+z^2}.$$

The *cylindrical coordinate system* is a 3D system that is simply an extrusion of polar coordinates. In this system, a point is identified by three coordinates (r, θ, z). If we consider polar coordinates to live in the xy-plane, then the third coordinate is simply the point's z-value.

A well-known conversion from spherical to cylindrical coordinates is the *Mercator projection.* For a given sphere (such as Earth, approximately), circumscribe a cylinder touching the sphere's equator. Now assign a point $\mathbf{p}_c$ on the cylinder to every point $\mathbf{p}_s$ on the sphere by intersecting the line through the sphere center and $\mathbf{p}_s$ with the cylinder, resulting in $\mathbf{p}_c$. Finally, cut the cylinder by a vertical line so it can be flattened out into a 2D map. Note that great circles on the sphere are mapped to straight lines on the 2D map. Furthermore, the Mercator projection has one big advantage: any angle formed by two great circles on the sphere is preserved by the corresponding straight lines of the 2D map. Such angle-preserving maps are called *conformal.* See Figure 3.6 for a 2D version of the Mercator projection.

Clearly, the Mercator projection does not work for the north and south poles. Another drawback is the severe distortion near the poles. In a Mercator projection, Greenland looks similar in size to Africa, whereas Greenland really has only 1/10 the area of Africa.

3.3 Case Study: UTM Coordinates

Any map converted from a sphere to a 2D plane will distort either angles, areas, or both. The Universal Transverse Mercator (UTM) coordinate system was designed to combat this problem in an empirically optimal way. First, ignore regions very close to the poles. Then, divide Earth into 60 longitudinal wedges, each spanning 6 degrees. Each wedge has a longitudinal center line at 3 degrees between its

Figure 3.6. Result of the Mercator projection.

borders. Now, take a transversal cylinder touching this center great circle,[1] and map the wedge to a 2D map using the Mercator projection principle. Repeat for all 60 wedges. Divide each wedge into 20 longitudinal pieces. The spherical wedge pieces map to their 2D counterparts with a tolerable distortion. Figure 3.7 shows how Earth is divided up into small pieces using UTM.

[1] All these cylinders are perpendicular to the one used in the Mercator projection, hence the name "transversal."

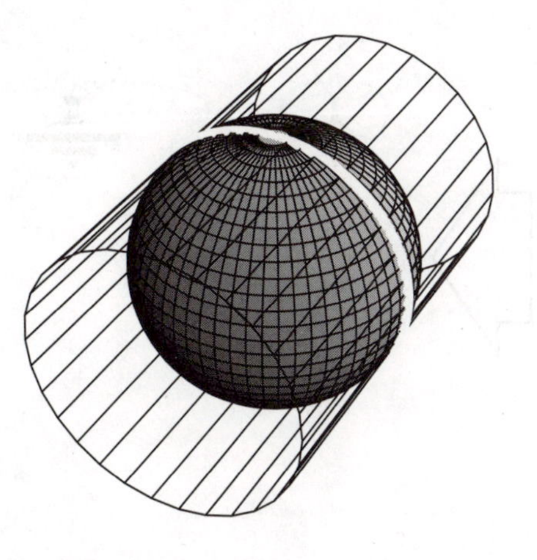

Figure 3.7. Geometry of the UTM coordinate system. One transversal cylinder and circle of contact are shown.

3.4 Local and Global Coordinates

How do we map points from one Cartesian coordinate system to another? The coordinate systems here are parallel to the 2D $\mathbf{o}, \mathbf{e}_i$ system defined in (3.1).

Suppose you have received an illustration of a bee from a graphics designer to use in your manuscript on the communication habits of bees. As illustrated on the left of Figure 3.8, the designer has created the artwork in the 2D Cartesian coordinate system, and its extents are $[0, 1]$ on both axes.[2] You would like to place several bees in an illustration. This repeated placement of the bee is illustrated in the right of the figure.

Let's introduce some terminology and notation first. We call the coordinate system that the designer used the *local coordinate system*. As we know from Section 3.1, a coordinate system is defined by an origin and axes. The local system in this case is simply the $\mathbf{o}, \mathbf{e}_1, \mathbf{e}_2$ system as defined in (3.1). Another way of thinking about a coordinate system is that it is defined by a rectangle called a *minmax box*: one corner corresponds to the origin, and the adjacent edges correspond to the coordinate axes. It is aligned with the coordinate

[2]Square brackets are used to indicate an interval.

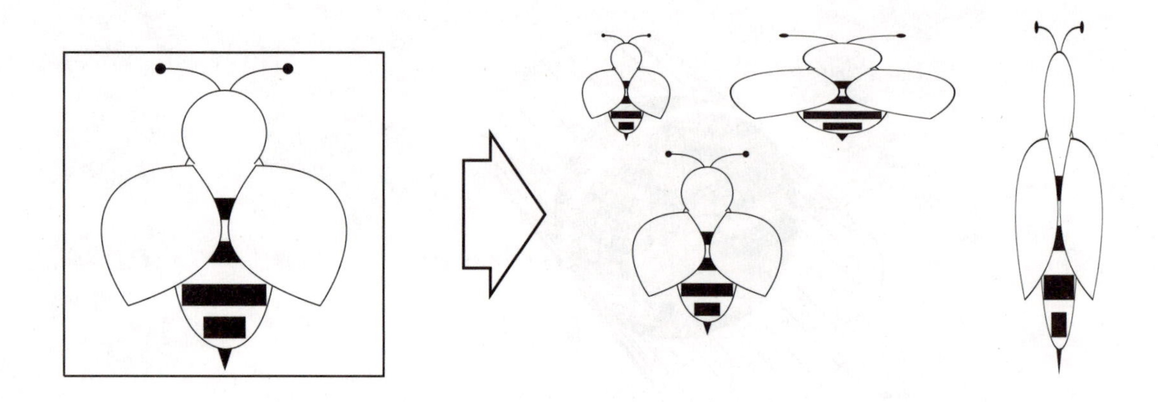

Figure 3.8. The bee design in a local coordinate system (left) and placement of several bees in global coordinate systems (right).

axes, and normally it encloses the extents of the geometry (the bee). The minmax box for the bee example is illustrated in Figure 3.8 (left): the lower-left point has coordinates $(0, 0)$ and the upper-right point has coordinates $(1, 1)$. Let (u_1, u_2) be a point in the local system. The local coordinates are also known as *parameters*.

The area on the page where you will place a copy of the bee will define a *global coordinate system*. Let's define this system by the minmax box with extents $(\min_1, \min_2)$ and $(\max_1, \max_2)$. Let (x_1, x_2) be a point in the global system. This rectangle is also called a *target box*.

Now we want to determine how to map a point (u_1, u_2) on the bee in the local system to a point (x_1, x_2) in the target box. Corresponding corners of the minmax boxes should map to each other. We want to preserve the ratio of a given coordinate with respect to its extents. If $u_1 = 1/2$ in the local system, then it should result in $x_1 = (1/2)\min_1 + (1/2)\max_1$ in the global system. To achieve this, we equate the following ratios:

$$\frac{u_1 - 0}{1 - 0} = \frac{x_1 - \min_1}{\max_1 - \min_1} \tag{3.3}$$

$$\frac{u_2 - 0}{1 - 0} = \frac{x_2 - \min_2}{\max_2 - \min_2}. \tag{3.4}$$

Thus, the corresponding formulas for x_1 and x_2 are quite simple:

$$x_1 = (1 - u_1)\mathrm{min}_1 + u_1\mathrm{max}_1, \tag{3.5}$$

$$x_2 = (1 - u_2)\mathrm{min}_2 + u_2\mathrm{max}_2. \tag{3.6}$$

We say that the coordinates (u_1, u_2) are *mapped* to the coordinates (x_1, x_2). Let's check this for the corners that define the extents. For $(u_1, u_2) = (0, 0)$ we map to

$$x_1 = (1 - 0) \cdot \mathrm{min}_1 + 0 \cdot \mathrm{max}_1 = \mathrm{min}_1,$$
$$x_2 = (1 - 0) \cdot \mathrm{min}_2 + 0 \cdot \mathrm{max}_2 = \mathrm{min}_2.$$

Similarly, the coordinates $(u_1, u_2) = (1, 1)$ in the local system must go to the coordinates $(x_1, x_2) = (\mathrm{max}_1, \mathrm{max}_2)$ in the global system. We obtain

$$x_1 = (1 - 1) \cdot \mathrm{min}_1 + 1 \cdot \mathrm{max}_1 = \mathrm{max}_1,$$
$$x_2 = (1 - 1) \cdot \mathrm{min}_2 + 1 \cdot \mathrm{max}_2 = \mathrm{max}_2.$$

A different way of writing (3.5) and (3.6) is as follows: Define $\Delta_1 = \mathrm{max}_1 - \mathrm{min}_1$ and $\Delta_2 = \mathrm{max}_2 - \mathrm{min}_2$. Now we have

$$x_1 = \mathrm{min}_1 + u_1\Delta_1, \tag{3.7}$$

$$x_2 = \mathrm{min}_2 + u_2\Delta_2. \tag{3.8}$$

A note of caution: if the target box is not a square, then the object from the local system will be distorted. We see this in two of the four bees on the right side of Figure 3.8. In general, if $\Delta_1 > 1$, then the object will be stretched in the $\mathbf{e}_1$-direction, and if $0 < \Delta_1 < 1$ it will be shrunk. The case of max_1 smaller than min_1 is not often encountered: it would result in a reversal of the object in the $\mathbf{e}_1$-direction. The same applies, of course, to the $\mathbf{e}_2$-direction. This change of shape of the object is characterized by the *aspect ratio*, which is the ratio of the width to the height, or Δ_1/Δ_2 for the target box. The aspect ratio in the local system is one.

We experience this "unit square to target box" mapping whenever we use a computer. Suppose we open a window to view an image. The image is stored in a local coordinate system; if it is stored with extents $(0, 0)$ and $(1, 1)$, then it utilizes *normalized coordinates*.

The target box is now given by the extents of the window, which are given in terms of *screen coordinates* and the image is mapped to it by using relationships (3.5) and (3.6). Screen coordinates are typically given in terms of *pixels*.[3]

How about the inverse problem: given coordinates (x_1, x_2) in the global system, what are its local (u_1, u_2) coordinates? The answer is relatively easy: compute u_1 from (3.7), and u_2 from (3.8), resulting in

$$u_1 = \frac{x_1 - \min_1}{\Delta_1}, \tag{3.9}$$

$$u_2 = \frac{x_2 - \min_2}{\Delta_2}. \tag{3.10}$$

It isn't necessary for the local minmax box to be a unit square. Simply modify (3.3) and (3.4) to reflect the desired ratios, and then reformulate (3.7) and (3.8).

This section has concentrated on 2D; however, 3D works just the same! We simply apply the same technique to the z-coordinate as well.

The mapping (scale and translation) from local to global coordinates is a special case of an *affine map*. An affine map is a combination of a linear map (e.g., scale or rotation) and a translation. Linear maps are investigated in more detail in Section 4.3. Affine maps allow us to rotate and move the bee, as illustrated in Figure 3.9.

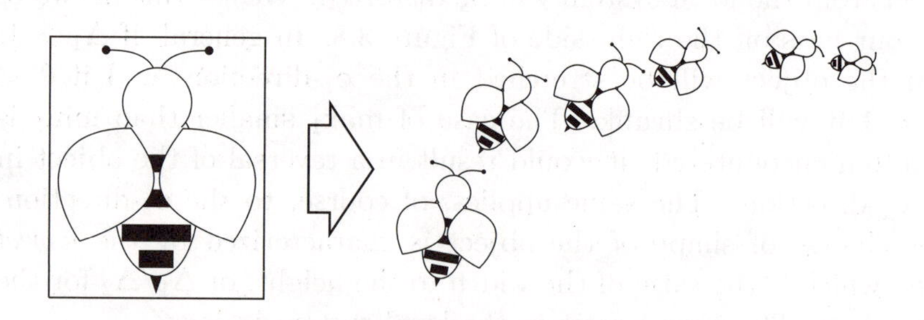

Figure 3.9. The bee design in a local coordinate system (left) and placement of several bees in general global coordinate systems (right).

[3]The term pixel is short for "picture element."

3.5 Homogeneous Coordinates

Most 2D and 3D geometry concepts are based on affine geometry. The maps ruling this kind of geometry, the *affine maps*, map parallel lines to parallel lines. This may sound appealing, but not all maps that we encounter are of this nature. A standard example is provided by looking at physical structures containing parallel lines (such as a straight railroad track). These lines do not appear parallel, however, when looking at the structure! This is due to the fact that our visual perception is based on 3D-to-2D perspective projections—and they do not behave like affine maps. We use *projective geometry* when dealing with this kind of phenomenon.

Let's consider Figure 3.10. The 3D point $\mathbf{x}$ is projected into the plane $z = 1$ by a ray through the origin—think of your as eye being at the origin and projecting into an image plane represented by the $z = 1$ plane. The projected point is $\mathbf{y}$; some simple math shows that

$$\mathbf{x} = \begin{bmatrix} x \\ y \\ z \end{bmatrix} \longrightarrow \begin{bmatrix} x/z \\ y/z \\ 1 \end{bmatrix} = \mathbf{y}.$$

Not just $\mathbf{x}$ is projected to $\mathbf{y}$, however—any multiple $\alpha\mathbf{x}$ projects to $\mathbf{y}$ as well. For this reason, we say that the 2D point $\mathbf{y}$[4] is

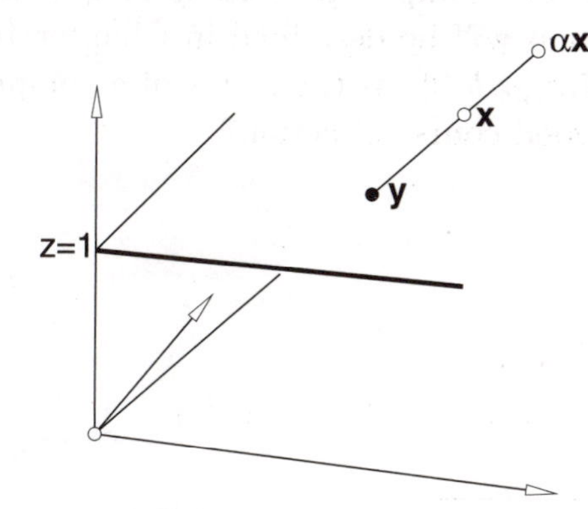

Figure 3.10. Geometry of homogeneous coordinates.

[4]This point does have a third coordinate, but this can easily be omitted.

represented by any point of the form $\alpha\mathbf{x}$. The collection of points $\alpha\mathbf{x}$ are the *homogeneous coordinates* of $\mathbf{y}$.

Using homogeneous coordinates simplifies many geometric computations. An extension leads to the concept of a *Clifford algebra*, an emerging field in computing and visualization.

3.6 Problems and Experiments

1. Given a 2D point $\mathbf{p}$, how does one find its polar coordinates?

2. Describe (in terms of spherical coordinates) how to find the shortest path between two locations.

3. An experiment with polar coordinates: Go to Google Earth. Find the North Pole and zoom in. You will see strange things happening.[5] Explain.

4. What type of projection does your favorite map application, such as Google Maps, use?

5. What is the aspect ratio a of a rectangle with a lower-left corner of $[1, 2]^{\mathrm{T}}$ and an upper-right corner of $[7, 4]^{\mathrm{T}}$?[6] Sketch this rectangle and a rectangle with aspect ratio $1/a$.

6. Suppose you are computing with homogeneous coordinates. (An application will be described in Chapter 16.) What happens if a point $[a, b, 0]^{\mathrm{T}}$ is the result of a computation? What would be a good course of action?

---

[5] This is as of the time of writing this book: Fall 2007.

[6] The superscript T indicates the transpose of the matrix, in which the rows are now columns, so $[1, 2]^{\mathrm{T}} = \begin{bmatrix} 1 \\ 2 \end{bmatrix}$. This notation is used commonly to save space in the text.

4

Background: Numerical Linear Algebra

A *matrix* is a rectangular array of numbers. Matrices appear as the following.

- *Digital images.* A computer screen consists of an array of pixels (about 1000×1000); each pixel is assigned a color.
- *The Google matrix.* The information about which webpage points to which other webpage is stored in a matrix of size 10 billion $\times$ 10 billion.
- *Transformation matrices.* The information about a 3D object's orientation is stored in a 3×3 matrix.

Matrices are important tools in scientific computing. They are dealt with in linear algebra; next, we give a brief review.

4.1 Linear Spaces and Vectors

First, we review the basic concepts of linear algebra. The most basic concept is that of a *linear space*, meaning a set in which certain simple (linear) operations are defined. A simple example is given by all 2D vectors. We denote vectors by boldface characters, such as $\mathbf{u}$

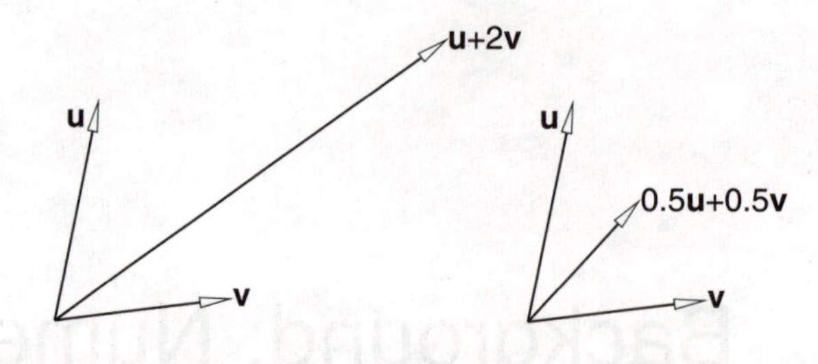

Figure 4.1. Linear combinations for 2D vectors.

or $\mathbf{v}$. Two-dimensional vectors have two components:[1]

$$\mathbf{u} = \begin{bmatrix} u_1 \\ u_2 \end{bmatrix}, \quad \mathbf{v} = \begin{bmatrix} v_1 \\ v_2 \end{bmatrix}.$$

The basic operation defined for vectors is the *linear combination*: given two vectors $\mathbf{u}$ and $\mathbf{v}$, we may combine them to obtain a third vector $\mathbf{w}$:

$$\mathbf{w} = s\mathbf{u} + t\mathbf{v}, \tag{4.1}$$

where s and t are some real numbers. The vector $\mathbf{w}$ is computed component by component:

$$\begin{bmatrix} w_1 \\ w_2 \end{bmatrix} = \begin{bmatrix} s \cdot u_1 + t \cdot v_1 \\ s \cdot u_2 + t \cdot v_2 \end{bmatrix}.$$

Two examples of linear combinations are shown in Figure 4.1.

An important vector for the special case in which $s = t = 0$ is the *zero vector*, denoted by a boldface zero, $\mathbf{0}$, which has zeroes for both components.

A 2D linear space exhibits the basic properties of all linear spaces: *a linear space is a set in which any two elements can be linearly combined and the result is again in the space.* Examples of linear spaces:

- The set of all digital images of the same size.
- The set of all differentiable functions.

[1] In the 2D case, we sometimes refer to these components as the x-component and the y-component.

- The set of all quadrilaterals in the plane.

It may be illuminating to consider an example that is *not* a linear space. Take the set of all 2D vectors $\mathbf{u} = [u_1, u_2]$ in which $u_1 \geq 0$. This is not a linear space. As an example, we take

$$\mathbf{u} = \begin{bmatrix} 1 \\ 0 \end{bmatrix} \quad \text{and} \quad \mathbf{v} = \begin{bmatrix} 0 \\ 1 \end{bmatrix}.$$

Now, picking $s = -1$ and $t = 1$ gives

$$\mathbf{w} = \begin{bmatrix} -1 \\ 1 \end{bmatrix},$$

which has a negative first component and hence is not in the space.

4.2 Linear Independence and Subspaces

We also consider vectors with more than two components; they are given by

$$\mathbf{u} = \begin{bmatrix} u_1 \\ u_2 \\ \vdots \\ u_n \end{bmatrix}.$$

All vectors of this form then constitute a linear space in which the basic linear operation is again given by (4.1), except that now it is involves n equations, not just two. The number n is referred to as the *dimension* of the space, and the space itself is denoted by $\mathbb{R}^n$.

There is more to be said about the concept of a dimension. It is closely linked to another basic concept, that of *linear independence.* Let's start with the 3D case, and consider three vectors,

$$\mathbf{u} = \begin{bmatrix} 1 \\ 1 \\ 1 \end{bmatrix}, \quad \mathbf{v} = \begin{bmatrix} 0 \\ 1 \\ 0 \end{bmatrix}, \quad \mathbf{w} = \begin{bmatrix} 2 \\ 1 \\ 2 \end{bmatrix},$$

for which $\mathbf{w} = 2\mathbf{u} - \mathbf{v}$. Thus, there exists a linear relationship among the vectors $\mathbf{u}, \mathbf{v}, \mathbf{w}$. We call such vectors *linearly dependent.*

Now, let's consider three other vectors,

$$\mathbf{u} = \begin{bmatrix} 1 \\ 0 \\ 0 \end{bmatrix}, \quad \mathbf{v} = \begin{bmatrix} 0 \\ 1 \\ 0 \end{bmatrix}, \quad \mathbf{w} = \begin{bmatrix} 0 \\ 0 \\ 1 \end{bmatrix}.$$

It should be clear that we cannot write $\mathbf{w}$ as a linear combination of $\mathbf{u}$ and $\mathbf{v}$ simply by looking at the third components. We define a set of vectors as *linearly independent* if no linear relation exists among them. Stated differently: in a set of linearly independent vectors, no vector can be written as a linear combination of the remaining ones. How do we know whether a given set of vectors is linearly dependent or independent? This question is dealt with in the context of linear systems in Chapter 5.

The *dimension* of a linear space is defined as the largest number of linearly independent vectors in it. This sounds abstract, but an example clarifies the definition: the dimension of $\mathbb{R}^2$ is 2, just as the name suggests. Why? Consider the two linearly independent vectors

$$\mathbf{u} = \begin{bmatrix} 1 \\ 0 \end{bmatrix}, \quad \mathbf{v} = \begin{bmatrix} 0 \\ 1 \end{bmatrix}.$$

Any vector $\mathbf{w}$ may be linearly expressed in terms of these two:

$$\mathbf{w} = \begin{bmatrix} w_1 \\ w_2 \end{bmatrix} = w_1\mathbf{u} + w_2\mathbf{v}.$$

Thus, although the two vectors $\mathbf{u}, \mathbf{v}$ are linearly independent, vectors $\mathbf{u}, \mathbf{v}, \mathbf{w}$ are linearly dependent.

This works for higher dimensions as well. The vectors

$$\begin{bmatrix} 1 \\ 0 \\ 0 \\ \vdots \\ 0 \end{bmatrix}, \quad \begin{bmatrix} 0 \\ 1 \\ 0 \\ \vdots \\ 0 \end{bmatrix}, \quad \begin{bmatrix} 0 \\ 0 \\ 1 \\ \vdots \\ 0 \end{bmatrix}, \quad \ldots, \quad \begin{bmatrix} 0 \\ 0 \\ 0 \\ \vdots \\ 1 \end{bmatrix}$$

form an independent set in $\mathbb{R}^n$.

If a linear space has dimension n and we have a set of n linearly independent vectors, then this set is referred to as a *basis* of the space. The reason for this term is that every element of the space can be expressed uniquely as a linear combination of the basis vectors. Finding that linear combination leads to linear systems of equations, which will be covered in Chapter 5.

To prepare for the next topic, consider the two linearly independent vectors

$$\mathbf{u} = \begin{bmatrix} 0 \\ 1 \\ 0 \end{bmatrix} \quad \text{and} \quad \mathbf{v} = \begin{bmatrix} 1 \\ 0 \\ 0 \end{bmatrix}.$$

They do not form a basis of $\mathbb{R}^3$, since, for example,

$$\mathbf{w} = \begin{bmatrix} 0 \\ 0 \\ 1 \end{bmatrix}$$

cannot be expressed as a linear combination of $\mathbf{u}$ and $\mathbf{v}$. However, they do generate a linear space: the set of all vectors $\mathbf{r}$ of the form

$$\mathbf{r} = s \cdot \mathbf{u} + t \cdot \mathbf{v} \tag{4.2}$$

has the required linear structure. Combining two vectors of the form (4.2), after some regrouping, again yields a linear combination of $\mathbf{u}$ and $\mathbf{v}$.

In general, assume we have a set of m linearly independent vectors $\mathbf{u}_1, \mathbf{u}_2, \ldots, \mathbf{u}_m$ in a linear space S of dimension n. Then all linear combinations that can be formed by the vectors $\mathbf{u}_i$ form a linear space U of dimension m. It is called a *subspace* of S, and U is *spanned* by the $\mathbf{u}_i$.

4.3 Linear Maps and Matrices

Finally, we need the concept of a *linear map*. We start with an example. In $\mathbb{R}^2$, let's stretch every vector $\mathbf{u}$ by a factor of 2 in the x-direction, and by a factor of 0.5 in the y-direction, obtaining *image vectors* $\mathbf{u}'$. We describe this operation—a linear map—as follows:

$$\begin{bmatrix} u_1' \\ u_2' \end{bmatrix} = \begin{bmatrix} 2u_1 \\ 0.5u_2 \end{bmatrix}.$$

If we give the linear map a name, say A, then we can write

$$\mathbf{u}' = A\mathbf{u}. \tag{4.3}$$

Just as vectors may have components or coordinates, map A may be given coordinates. It can then be written in *matrix form.* A matrix is a rectangular array A of real numbers $a_{i,j}$. Relation (4.3) involves the multiplication of a matrix A and a vector $\mathbf{u}$, which is defined as

$$A \cdot \mathbf{u} = \begin{bmatrix} a_{1,1} & a_{1,2} \\ a_{2,1} & a_{2,2} \end{bmatrix} \begin{bmatrix} u_1 \\ u_2 \end{bmatrix} = \begin{bmatrix} a_{1,1}u_1 + a_{1,2}u_2 \\ a_{2,1}u_1 + a_{2,2}u_2 \end{bmatrix}.$$

For our specific scaling example, we have

$$A \cdot \mathbf{u} = \begin{bmatrix} 2 & 0 \\ 0 & 0.5 \end{bmatrix} \begin{bmatrix} u_1 \\ u_2 \end{bmatrix} = \begin{bmatrix} 2u_1 \\ 0.5u_2 \end{bmatrix}.$$

Reflections may be obtained by scaling with negative factors. Linear maps can do more than just scale, however. Other linear maps involve rotations and shears. An example of a shear is given by

$$A \cdot \mathbf{u} = \begin{bmatrix} 1 & 1 \\ 0 & 1 \end{bmatrix} \begin{bmatrix} u_1 \\ u_2 \end{bmatrix} = \begin{bmatrix} u_1 + u_2 \\ u_2 \end{bmatrix}.$$

Its action is shown in Figure 4.2.

In general, linear maps take the form $\mathbf{v} = A\mathbf{u}$, where $\mathbf{v} \in \mathbb{R}^m$, $\mathbf{u} \in \mathbb{R}^n$, and A is a matrix with m rows and n columns. The ith element of $\mathbf{v}$, v_i, is

$$v_i = a_{i,1}u_1 + a_{i,2}u_2 + \ldots + a_{i,n}u_n; \quad i = 1, \ldots, m. \tag{4.4}$$

The ith element of $\mathbf{v}$ is obtained by using all elements of $\mathbf{u}$, but only the ith row of A. Expressions of the form (4.4) are called *dot products.* Thus, v_i is the dot product of A's ith row with $\mathbf{u}$.

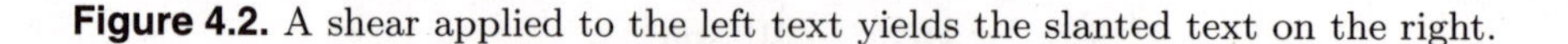

Figure 4.2. A shear applied to the left text yields the slanted text on the right.

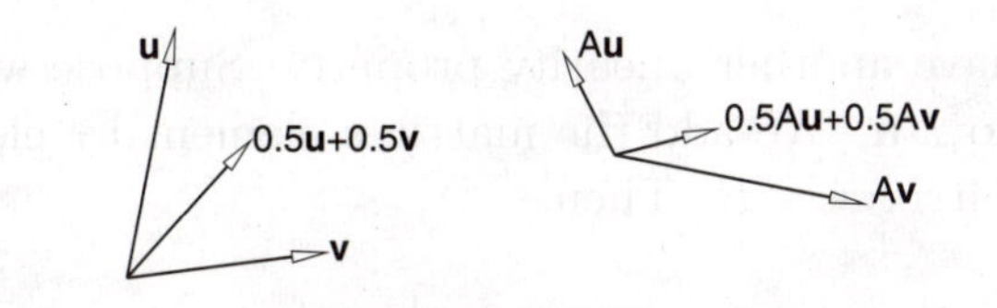

Figure 4.3. Three vectors (left) and their images under a linear map (right).

For example, a map from $\mathbb{R}^3$ to $\mathbb{R}^2$ might be

$$\begin{bmatrix} 1 & 0 & -1 \\ 1 & 2 & 1 \end{bmatrix} \begin{bmatrix} 1 \\ -1 \\ 0 \end{bmatrix} = \begin{bmatrix} 1 \\ -1 \end{bmatrix}.$$

For any matrix A and vectors $\mathbf{u}$, $\mathbf{v}$, we have the relation

$$A(s\mathbf{u} + t\mathbf{v}) = sA\mathbf{u} + tA\mathbf{v}. \tag{4.5}$$

This *linearity property* is extremely important: it states that linear maps preserve linear relationships. For example, the average $0.5\mathbf{u} + 0.5\mathbf{v}$ will be mapped to the average of $A\mathbf{u}$ and $A\mathbf{v}$, as shown in Figure 4.3.

Other, so-called nonlinear maps, do not enjoy the linearity property. For example, let's take the map

$$\begin{bmatrix} u_1 \\ u_2 \end{bmatrix} \rightarrow \begin{bmatrix} u_1 \\ u_2^2 \end{bmatrix}.$$

Then

$$\mathbf{u}_1 = \begin{bmatrix} 1 \\ 2 \end{bmatrix} \rightarrow \begin{bmatrix} 1 \\ 4 \end{bmatrix} \quad \text{and} \quad \mathbf{u}_2 = \begin{bmatrix} 2 \\ 4 \end{bmatrix} \rightarrow \begin{bmatrix} 2 \\ 16 \end{bmatrix}$$

but the average of $\mathbf{u}_1$ and $\mathbf{u}_2$:

$$\begin{bmatrix} 1.5 \\ 3 \end{bmatrix}$$

is not mapped to the average of the images, which is given by

$$\begin{bmatrix} 1.5 \\ 10 \end{bmatrix}.$$

Instead it is mapped to $[1.5, 9]^{\mathrm{T}}$.

Matrices have another linearity property. Suppose we map $\mathbf{u}$ to $A\mathbf{u}$ and also to $B\mathbf{u}$. We add the matrices element by element, thus obtaining a matrix $A + B$. Then

$$[A + B]\mathbf{u} = A\mathbf{u} + B\mathbf{u}.$$

Similarly, we may multiply A by a scalar s, obtaining sA, by multiplying every element of A by s.

An important special matrix is the *identity matrix*, typically denoted by I. Its diagonal elements (those having two identical subscripts) are 1, and all of the other elements are 0. The 3×3 identity matrix is

$$I = \begin{bmatrix} 1 & 0 & 0 \\ 0 & 1 & 0 \\ 0 & 0 & 1 \end{bmatrix}.$$

The reason for the term "identity" is simple: for every vector $\mathbf{u}$, $I\mathbf{u} = \mathbf{u}$.

Further, actions of linear maps may be combined. Let

$$\mathbf{v} = A\mathbf{u} \quad \text{and} \quad \mathbf{w} = B\mathbf{v}.$$

Then

$$\mathbf{w} = B \cdot A \cdot \mathbf{u}$$

and the *matrix product* $C = BA$ is defined as follows. Each element $c_{i,j}$ is defined as the dot product of the ith row of B and the jth column of A. For that to work, B needs to have as many columns as A has rows. Figure 4.4 shows an easy way to arrange the matrices for hand computation. An equation for $c_{i,j}$ thus has exactly the same structure as (4.4):

$$c_{i,j} = b_{i,1}a_{1,j} + b_{i,2}a_{2,j} + \ldots + b_{i,n}a_{n,j}.$$

Let's look at a numerical example:

$$\begin{bmatrix} 1 & 0 & -1 \\ 1 & 2 & 1 \end{bmatrix} \begin{bmatrix} 1 & 2 \\ 2 & -3 \\ -1 & 1 \end{bmatrix} = \begin{bmatrix} 2 & 1 \\ 4 & -3 \end{bmatrix}.$$

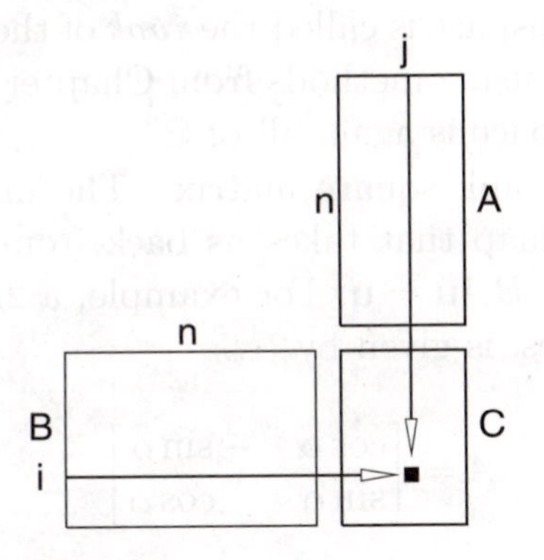

Figure 4.4. Matrix multiplication.

The product of two matrices may be written as the sum of very simple matrices. Let $\mathbf{b}_r$ be the rth column of B and let $\mathbf{a}_r^{\mathrm{T}}$ the rth row of A. Then,

$$BA = \mathbf{b}_1\mathbf{a}_1^{\mathrm{T}} + \ldots + \mathbf{b}_n\mathbf{a}_n^{\mathrm{T}}. \tag{4.6}$$

For the preceding example, we have

$$\begin{aligned} BA = \begin{bmatrix} 2 & 1 \\ 4 & -3 \end{bmatrix} &= \begin{bmatrix} 1 \\ 1 \end{bmatrix} \begin{bmatrix} 1 & 2 \end{bmatrix} + \begin{bmatrix} 0 \\ 2 \end{bmatrix} \begin{bmatrix} 2 & -3 \end{bmatrix} + \begin{bmatrix} -1 \\ 1 \end{bmatrix} \begin{bmatrix} -1 & 1 \end{bmatrix} \\ &= \begin{bmatrix} 1 & 2 \\ 1 & 2 \end{bmatrix} + \begin{bmatrix} 0 & 0 \\ 4 & -6 \end{bmatrix} + \begin{bmatrix} 1 & -1 \\ -1 & 1 \end{bmatrix} = \begin{bmatrix} 2 & 1 \\ 4 & -3 \end{bmatrix}. \end{aligned}$$

Each of the matrices $\mathbf{b}_i\mathbf{a}_i^{\mathrm{T}}$ is called a *dyadic matrix.* This unusual way of writing matrix multiplication will be used for a method called principal components analysis (PCA) in Section 6.9.

When multiplying matrices, *order matters.* That is to say, in general

$$AB \neq BA.$$

In the above example, the matrix dimensions would not match if we switched the order, but even if they match, order does matter.

We next discuss the concept of the *rank* of a matrix. Instead of striving for full generality, we will restrict ourselves to square matrices, mapping $\mathbb{R}^n$ to $\mathbb{R}^n$. Let $\mathbf{u}_1, \ldots, \mathbf{u}_n$ be a set of n linearly independent vectors. Let their images under some linear map A be $A\mathbf{u}_1, \ldots, A\mathbf{u}_n$. These image vectors span a subspace of $\mathbb{R}^n$. The dimension of this subspace is called the *rank* of the matrix. The rank can be computed by using methods from Chapter 5. The matrix has *full rank* if the subspace is again all of $\mathbb{R}^n$.

Let A be a full rank square matrix. The image of a vector $\mathbf{u}$ is $A\mathbf{u}$. We want a map that takes us back from $A\mathbf{u}$ to $\mathbf{u}$. Such a map B is defined by $BA\mathbf{u} = \mathbf{u}$. For example, a 2D rotation matrix, rotating by α degrees, is given by

$$A = \begin{bmatrix} \cos\alpha & -\sin\alpha \\ \sin\alpha & \cos\alpha \end{bmatrix}.$$

Then matrix B would be a matrix that rotates by $-\alpha$ degrees:

$$B = \begin{bmatrix} \cos\alpha & \sin\alpha \\ -\sin\alpha & \cos\alpha \end{bmatrix}.$$

Thus, for any $\mathbf{u}$, $BA\mathbf{u} = \mathbf{u}$. We can also verify that $BA = I$.

In general, the matrix that "undoes" the action of a matrix A is called A's *inverse* and denoted by A^{-1}. Clearly,

$$A^{-1}A = I. \tag{4.7}$$

Further, we have the obvious identity

$$A^{-1^{-1}} = A$$

and the not-so-obvious identity

$$(AB)^{-1} = B^{-1}A^{-1}.$$

The computation of the inverse is discussed in Chapter 5. It is important to note here that not all square matrices have an inverse: this is true only if they are of full rank. Non-full-rank matrices are called *singular*. They can map different vectors to the same image vectors as the following example illustrates. Let

$$A = \begin{bmatrix} 1 & 2 & 3 \\ 2 & 4 & 6 \\ 3 & 6 & 9 \end{bmatrix}$$

be a 3×3 matrix of rank one. (Observe that the last row is the sum of the first and second rows and the second row is a multiple of the first one.) We can map two different vectors using A:

$$\begin{bmatrix} 1 & 2 & 3 \\ 2 & 4 & 6 \\ 3 & 6 & 9 \end{bmatrix} \begin{bmatrix} -4 \\ 2 \\ 2 \end{bmatrix} = \begin{bmatrix} 6 \\ 12 \\ 18 \end{bmatrix} \quad \text{and} \quad \begin{bmatrix} 1 & 2 & 3 \\ 2 & 4 & 6 \\ 3 & 6 & 9 \end{bmatrix} \begin{bmatrix} 0 \\ 3 \\ 0 \end{bmatrix} = \begin{bmatrix} 6 \\ 12 \\ 18 \end{bmatrix}.$$

Two different vectors, same result! Clearly, one cannot say that $[6, 12, 18]^{\mathrm{T}}$ is the image of *one* vector. Hence, A is singular.

4.4 Lengths and Volumes

What is the length of a vector? In 2D, this is simple: a vector $\mathbf{u}$ with components u_1 and u_2 has length $\sqrt{u_1^2 + u_2^2}$. We denote this length by $\|\mathbf{u}\|$. In general, an n-dimensional vector $\mathbf{u}$ has length

$$\|\mathbf{u}\| = \sqrt{u_1^2 + \ldots + u_n^2}.$$

The term under the square root may be written more elegantly. We can use the *dot product* of $\mathbf{u}^{\mathrm{T}}$ and $\mathbf{u}$ to find the length:

$$\|\mathbf{u}\| = \sqrt{\mathbf{u}^{\mathrm{T}}\mathbf{u}}.$$

A vector is called a *unit vector* if $\|\mathbf{u}\| = 1$. In $\mathbb{R}^n$, all unit vectors form the n-dimensional *unit sphere.*

As mentioned previously, the general definition of a dot product involves two vectors $\mathbf{u}$ and $\mathbf{v}$ and is defined as

$$\mathbf{u} \cdot \mathbf{v} = u_1 v_1 + \ldots + u_n v_n. \tag{4.8}$$

Dot products (also called scalar products) are important because they describe the geometry of the two vectors. Two vectors (in any dimension) form an angle. The cosine of this angle is computed by using the dot product $\mathbf{u} \cdot \mathbf{v}$:

$$\cos(\mathbf{u}, \mathbf{v}) = \frac{\mathbf{u} \cdot \mathbf{v}}{\|\mathbf{u}\|\|\mathbf{v}\|}. \tag{4.9}$$

Since two perpendicular vectors enclose an angle of 90° (or $\pi/2$), we have a simple condition: *two vectors* $\mathbf{u}, \mathbf{v}$ *are perpendicular if* $\mathbf{u} \cdot \mathbf{v} = 0$.

For example, let

$$\mathbf{u} = \begin{bmatrix} 1 \\ 3 \end{bmatrix} \quad \text{and} \quad \mathbf{v} = \begin{bmatrix} -6 \\ 2 \end{bmatrix}.$$

Check for yourself that $\mathbf{u} \cdot \mathbf{v} = 0$ and sketch the vectors!

Perpendicular vectors turn out to have many desirable properties. Generally speaking, they make computations more stable.

Let us consider n unit vectors $\mathbf{u}_1, \ldots, \mathbf{u}_n$ in $\mathbb{R}^n$. They are *mutually orthogonal* if any two distinct ones are orthogonal (which is another word for perpendicular):

$$\mathbf{u}_i \cdot \mathbf{u}_j = \begin{cases} 1 & \text{if } i = j, \\ 0 & \text{if } i \neq j. \end{cases} \tag{4.10}$$

Many times, this is abbreviated to

$$\mathbf{u}_i \cdot \mathbf{v}_j = \delta_{i,j},$$

where $\delta_{i,j}$ is the *Kronecker delta.*

We recall that for a matrix A, the *transpose* is denoted by A^{T}; its elements $a'_{i,j}$ are defined by

$$a'_{i,j} = a_{j,i},$$

or row i of A^{T} is column i of A. We immediately have

$$A^{\mathrm{T}\mathrm{T}} = A.$$

Now we let U be the matrix whose columns are n unit vectors $\mathbf{u}_1, \ldots, \mathbf{u}_n$. If the $\mathbf{u}_i$ are mutually orthogonal, then U is called *orthonormal.* Its defining property is

$$U^{\mathrm{T}}U = I. \tag{4.11}$$

Notice that this equation is just a matrix formulation of (4.10). Because of (4.7), we also have

$$U^{-1} = U^{\mathrm{T}}. \tag{4.12}$$

The most obvious example of an orthonormal matrix is the identity matrix I. Another example is given by 2D rotation matrices.

Whereas the inverse of a general matrix is not easy to compute, it is highly trivial for orthonormal matrices!

There is another type of matrix for which we can compute the inverse without any problem. That is the *diagonal matrix* D, which is given by

$$D = \begin{bmatrix} d_1 & 0 & \cdots & 0 \\ 0 & d_2 & & \\ \vdots & & \ddots & \\ 0 & & & d_n \end{bmatrix}. \tag{4.13}$$

It has (nonzero) entries only on the diagonal and zero entries everywhere else. Its inverse is

$$D^{-1} = \begin{bmatrix} 1/d_1 & 0 & \cdots & 0 \\ 0 & 1/d_2 & & \\ \vdots & & \ddots & \\ 0 & & & 1/d_n \end{bmatrix}, \tag{4.14}$$

and we see that $DD^{-1} = I$.

Let us now turn to the *volume* formed by a set of vectors. As an example, the two unit vectors $[1, 0]^{\mathrm{T}}, [0, 1]^{\mathrm{T}}$ form a square whose area (or 2D volume) is 1. We generalize this to say that the volume generated by all column vectors of I (for an arbitrary dimension n) is 1.

If we scaled the above 2D vectors so they become $[r, 0]^{\mathrm{T}}, [0, s]^{\mathrm{T}}$, then they form a rectangle with area $r \cdot s$. Again, we generalize to $\mathbb{R}^n$ and consider the volume generated by the column vectors of the diagonal matrix D in (4.13). It is given by $d_1 \cdot d_2 \cdot \ldots \cdot d_n$. We denote this volume by $|D|$:

$$|D| = d_1 \cdot d_2 \cdot \ldots \cdot d_n.$$

The volume $|D|$ is also referred to as the *determinant* of D, or $\det(D)$.

If U is an orthonormal matrix, then its column vectors can be mapped to those of I by a rotation. Since rotations do not change volumes, we have $|U| = 1$.

If a matrix A is singular, then $|A| = 0$. A simple example is a diagonal matrix of the form (4.13) where one of the d_i is zero.

For a general square matrix A, the computation of $|A|$ will be discussed in Sections 5.3 and 6.6.

4.5 Summary of Matrix Rules

For the following rules, it is expected that the involved matrices have compatible formats.

Commutative law for addition	$A+B=B+A$
No commutative law for multiplication	$AB \neq BA$
Distributive laws	$A(B+C)=AB+AC$
	$(B+C)A=BA+CA$
Rules for transpose matrices	$(A+B)^{\mathrm{T}}=A^{\mathrm{T}}+B^{\mathrm{T}}$
	$(AB)^{\mathrm{T}}=B^{\mathrm{T}}A^{\mathrm{T}}$
	${A^{-1}}^{\mathrm{T}}={A^{\mathrm{T}}}^{-1}$
Rules for inverse matrices:	$(AB)^{-1}=B^{-1}A^{-1}$
	$(A+B)^{-1} \neq A^{-1}+B^{-1}$

4.6 Problems and Experiments

1. In Section 4.1 we stated that the set of all digital images of the same size form a linear space. In order for this set to be a linear space, it needs to be equipped with the concept of linear combinations. Provide an appropriate definition of linear combinations.

2. Let a square matrix be populated by random real numbers. What is its chance of being singular? (Before considering a general discussion, concentrate on the 2×2 case.)

3. Find two 2×2 matrices A and B for which $AB \neq BA$.

4. Equation (4.6) makes use of dyadic matrices. What is the rank of a dyadic matrix?

5

Solving Linear Systems

This chapter introduces basic methods for solving a linear system of equations. Gauss elimination is the most fundamental method; however, it is not appropriate for all problems. Iterative methods, such as Gauss-Jacobi and Gauss-Seidel, are introduced as well. Stability and convergence concepts are an important part of these iterative methods. If more information (data points) for a problem is given than is needed for a solution, this leads to an overdetermined system of equations. Many solutions are possible; an optimal solution, called least squares, is described here. Three case studies motivate the ideas in this chapter.

5.1 Case Study: Mixing Chemicals

A chemist has, from a past experiment, three containers, each containing a mixture of three substances A, B, and C. For concreteness, let's assume the following amounts:

	Container 1	Container 2	Container 3
A	40 g	40 g	70 g
B	30 g	80 g	60 g
C	60 g	10 g	30 g

The chemist wants to obtain another mixture by properly combining parts of the contents of each container. She wants this new mixture to consist of 50 grams each of A, B, and C. The question thus is, what fractions f_1, f_2, f_3 does she need to take from each container? By writing this down in the form of a system of equations, we get:

$$\begin{aligned} 40f_1 + 40f_2 + 70f_3 &= 50, \\ 30f_1 + 80f_2 + 60f_3 &= 50, \\ 60f_1 + 10f_2 + 30f_3 &= 50. \end{aligned}$$

These are three equations for three unknowns, f_1, f_2, f_3. In matrix form, they become

$$\begin{bmatrix} 40 & 40 & 70 \\ 30 & 80 & 60 \\ 60 & 10 & 30 \end{bmatrix} \begin{bmatrix} f_1 \\ f_2 \\ f_3 \end{bmatrix} = \begin{bmatrix} 50 \\ 50 \\ 50 \end{bmatrix}$$

By using a software package, we can quickly solve this for the required fractions:

$$\begin{bmatrix} f_1 \\ f_2 \\ f_3 \end{bmatrix} = \begin{bmatrix} 0.70 \\ 0.22 \\ 0.19 \end{bmatrix},$$

meaning we take 70% of the contents of container 1, 22% of container 2, and 19% of container 3.

5.2 Linear System Basics

A linear system of equations for unknowns $u_1, \ldots, u_n$ is given by

$$\begin{bmatrix} a_{1,1} & \cdots & a_{1,n} \\ \vdots & & \vdots \\ a_{n,1} & \cdots & a_{n,n} \end{bmatrix} \begin{bmatrix} u_1 \\ \vdots \\ u_n \end{bmatrix} = \begin{bmatrix} b_1 \\ \vdots \\ b_n \end{bmatrix}. \tag{5.1}$$

If convenient, it may be abbreviated to

$$A\mathbf{u} = \mathbf{b}.$$

If a solution $\mathbf{u}$ exists (which is not always the case), it may be written as

$$\mathbf{u} = A^{-1}\mathbf{b},$$

with A^{-1} being A's inverse. However, computing the inverse (using some package) as a means for solving a linear system is highly inefficient, which does not matter for 3×3 linear systems, but it does for $30,000 \times 30,000$ ones! The methods below are far more efficient.

If you took an elementary linear algebra class, you probably saw a method, known as *Cramer's rule*, for explicitly writing down the solution. We will not discuss this method here—although it has theoretical value, it is horribly inefficient for practical computing.

5.3 Gauss Elimination

The oldest systematic method for solving a linear system goes back to C. F. Gauss, who developed the following tool for finding the exact location of a newly discovered comet. The main "trick" is to cleverly multiply equations by factors and to then add them to each other. For example, of the two equations

$$\begin{aligned} 2u_1 - 2u_2 &= 10, \\ 6u_1 + 4u_2 &= 10, \end{aligned}$$

we may multiply both the left- and right-hand sides of the first one by -3 and then add the result to the second one (remembering that this is a valid operation for handling equations):

$$\begin{aligned} 2u_1 - 2u_2 &= 10 \\ 10u_2 &= -20. \end{aligned}$$

This linear system is equivalent to the first one, but now the second equation does not contain a u_1-term anymore, and we readily find $u_2 = -2$. Inserting this value into the original first equation gives $u_1 = 3$.

In matrix notation, we transformed the system

$$\begin{bmatrix} 2 & -2 \\ 6 & 4 \end{bmatrix} \begin{bmatrix} u_1 \\ u_2 \end{bmatrix} = \begin{bmatrix} 10 \\ 10 \end{bmatrix}$$

into the equivalent one

$$\begin{bmatrix} 2 & -2 \\ 0 & 10 \end{bmatrix} \begin{bmatrix} u_1 \\ u_2 \end{bmatrix} = \begin{bmatrix} 10 \\ -20 \end{bmatrix}.$$

In general, Gauss elimination transforms a given linear system into an equivalent one that can easily be solved, starting from the last unknown. We schematically show the transformation process for a 4×4 system—stars stand for entries that are nonzero in general:

$$\begin{bmatrix} \star & \star & \star & \star \\ \star & \star & \star & \star \\ \star & \star & \star & \star \\ \star & \star & \star & \star \end{bmatrix} \Rightarrow \begin{bmatrix} \star & \star & \star & \star \\ 0 & \star & \star & \star \\ 0 & \star & \star & \star \\ 0 & \star & \star & \star \end{bmatrix} \Rightarrow \begin{bmatrix} \star & \star & \star & \star \\ 0 & \star & \star & \star \\ 0 & 0 & \star & \star \\ 0 & 0 & \star & \star \end{bmatrix} \Rightarrow \begin{bmatrix} \star & \star & \star & \star \\ 0 & \star & \star & \star \\ 0 & 0 & \star & \star \\ 0 & 0 & 0 & \star \end{bmatrix}.$$

In actual implementations, care must be taken not to have zero entries on the diagonal—this would corrupt the process![1] If we cannot avoid ending up with a bottom diagonal element equal to zero, then the system has no solution. This is demonstrated in the following example.

The system

$$\begin{bmatrix} 2 & -2 & 0 \\ 6 & 4 & -1 \\ 8 & 2 & -1 \end{bmatrix} \begin{bmatrix} u_1 \\ u_2 \\ u_3 \end{bmatrix} = \begin{bmatrix} 10 \\ 10 \\ -10 \end{bmatrix}$$

is transformed (by multiplying row 1 by -3 and adding the result to row 2; then by multiplying row 1 by -4 and adding the result to row 3) to

$$\begin{bmatrix} 2 & -2 & 0 \\ 0 & 10 & -1 \\ 0 & 10 & -1 \end{bmatrix} \begin{bmatrix} u_1 \\ u_2 \\ u_3 \end{bmatrix} = \begin{bmatrix} 10 \\ -20 \\ -50 \end{bmatrix}.$$

Without performing the last elimination step, we can see that the last two equations contradict each other—there is no solution to this linear system.

Gauss elimination transforms a linear system $A\mathbf{u} = \mathbf{b}$ into an equivalent one

$$T\mathbf{u} = \mathbf{c}$$

in which T is an *upper triangular matrix.* The linear system has a solution if the bottom right element $t_{n,n}$ of T is not zero.

[1]The process of avoiding this is called *pivoting* and involves interchanging columns of the coefficient matrix.

The product

$$|A| = t_{1,1} \cdot t_{2,2} \cdot \ldots \cdot t_{n,n}$$

is known as A's *determinant*. Hence a linear system has a solution only if $|A| \neq 0$.

If $|A|$ is small (10^{-10}, say), then the solution to the linear system is likely to be afflicted by numerical error. See Section 6.6 on a method called singular value decomposition (SVD) for more details.

5.4 Stability

Consider the linear system

$$\begin{bmatrix} 0.8 & 0.1 & 0.1 \\ 0.7 & 0.24 & 0.06 \\ 0.6 & 0.4 & 0 \end{bmatrix} \begin{bmatrix} u_1 \\ u_2 \\ u_3 \end{bmatrix} = \begin{bmatrix} 9.49 \\ 8.856 \\ 8.16 \end{bmatrix}, \tag{5.2}$$

which has the solution

$$\mathbf{u} = \begin{bmatrix} 10.2 \\ 5.1 \\ 8.2 \end{bmatrix}.$$

Now consider the linear system

$$\begin{bmatrix} 0.8 & 0.1 & 0.1 \\ 0.7 & 0.24 & 0.06 \\ 0.6 & 0.4 & 0 \end{bmatrix} \begin{bmatrix} u_1 \\ u_2 \\ u_3 \end{bmatrix} = \begin{bmatrix} 9.69 \\ 8.856 \\ 8.16 \end{bmatrix}, \tag{5.3}$$

which has the solution

$$\mathbf{u} = \begin{bmatrix} 12.6 \\ 1.5 \\ -5.4 \end{bmatrix}.$$

We need to look a little closer to see why there is a surprise here. The two linear systems are almost identical—the only difference is a change by 2% in the first entry of the right-hand side of the second system. But surprisingly, the resulting solution is not even close to the original one!

We observe that a tiny change to the input data (the right-hand side) to a linear system may result in drastic changes in the solution. Such problems are called *unstable* or *ill-conditioned.*

The reason for such behavior will be discussed in Section 6.6 on SVD; it is related to the so-called condition number of a matrix. The lesson to be learned from this is that blindly using a linear system solver may lead to unreliable results.

Unstable behavior of a problem has nothing to do with the performance of a particular algorithm for its solution. No matter what algorithm we pick, slight changes, such as roundoff, to input data may result in large changes to output data.

5.5 Vector Norms and Sequences

We are all familiar with sequences of real numbers such as

$$1, \frac{1}{2}, \frac{1}{4}, \frac{1}{8}, \ldots$$

or

$$1, 2, 4, 8, \ldots.$$

The first of these has the *limit* 0, whereas the second one does not have a limit. One way of saying a sequence of real numbers a_i has a limit a is that beyond some index i, all a_i differ from the limit a by an arbitrarily small amount ϵ.

Vector sequences

$$\mathbf{v}^{(0)}, \mathbf{v}^{(1)}, \mathbf{v}^{(2)}, \ldots$$

are not all that different. Here we say the sequence has a limit $\mathbf{v}$ if from some index i on, the distance of any $\mathbf{v}^{(i)}$ from $\mathbf{v}$ is smaller than an arbitrarily small amount ϵ. By "distance" of two vectors, we are referring to the usual *Euclidean norm*: if $\mathbf{w} = \mathbf{v}^{(i+1)} - \mathbf{v}^{(i)}$, then the length or magnitude of $\mathbf{w}$ is given by

$$\|\mathbf{w}\| = \sqrt{\mathbf{w} \cdot \mathbf{w}}.$$

Figure 5.1 presents an example. The Euclidean norm is also known as the L^2 *norm*, and sometimes it is written as $\|\mathbf{w}\|_2$.

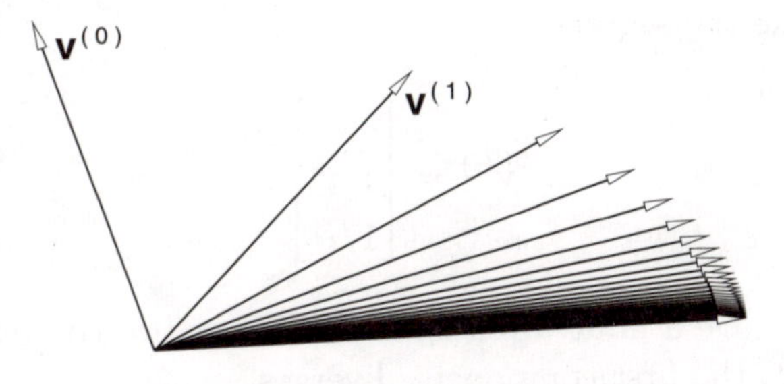

Figure 5.1. Vector sequences: a sequence that converges.

Different measures for vectors exist; for example, the *Manhattan norm* or L^1 *norm* is

$$\|\mathbf{w}\|_1 = |w_1| + \ldots + |w_n|.$$

We could define other norms as well; (nonnegative) vector norms must satisfy the following criteria:

- $\|\mathbf{w}\| > 0$ when $\mathbf{w} \neq \mathbf{0}$,
- $\|\mathbf{w}\| = 0$ when $\mathbf{w} = \mathbf{0}$,
- $\|c\mathbf{w}\| = |c|\|\mathbf{w}\|$ for any scalar c,
- $\|\mathbf{v} + \mathbf{w}\| \leq \|\mathbf{v}\| + \|\mathbf{w}\|$ (known as the triangle inequality).

For our purposes, the Euclidean norm will do, so unless otherwise noted, this the norm of choice and the 2 subscript will be omitted.

Let's take a vector sequence given by

$$\mathbf{v}^{(i)} = \begin{bmatrix} 1/i \\ 1/i^2 \\ 1/i^3 \end{bmatrix}.$$

This sequence has the limit

$$\mathbf{v} = \begin{bmatrix} 0 \\ 0 \\ 0 \end{bmatrix}.$$

Now we take the sequence

$$\mathbf{v}^{(i)} = \begin{bmatrix} i \\ 1/i^2 \\ 1/i^3 \end{bmatrix}.$$

It does not have a limit: even though the last two components each have a limit, the first component diverges.

5.6 Iterative System Solvers

In applications such as finite element methods (FEMs) in the context of the solution of fluid flow problems, scientists are faced with linear systems with many thousands of equations. Gauss elimination would work, but would be far too slow. Typically, huge linear systems have one advantage: the coefficient matrix is *sparse*, meaning it has only a few (such as ten) nonzero entries per row. Thus a 100,000 × 100,000 system would only have 1,000,000 nonzero entries, compared to 10,000,000,000 matrix elements! In these cases, one does not store the whole matrix, but only its nonzero entries, together with their i, j locations. Sections 5.7 and 11.6 give instances in which sparse matrices naturally arise in the solution of partial differential equations. The solution to such systems is typically obtained by *iterative methods*, which we discuss next.

We start with a simple (nonsparse) example. Let the system[2] be given by

$$\begin{bmatrix} 4 & 1 & 0 \\ 2 & 5 & 1 \\ -1 & 2 & 4 \end{bmatrix} \begin{bmatrix} u_1 \\ u_2 \\ u_3 \end{bmatrix} = \begin{bmatrix} 1 \\ 0 \\ 3 \end{bmatrix}.$$

By using an iterative method, we start from a guess for the solution and then refine it until we find the solution. Let's take

$$\mathbf{u}^{(1)} = \begin{bmatrix} u_1^{(1)} \\ u_2^{(1)} \\ u_3^{(1)} \end{bmatrix} = \begin{bmatrix} 1 \\ 1 \\ 1 \end{bmatrix}$$

[2]This example is taken from Johnson and Riess [10].

for our first guess. We note that it clearly is not the solution to our system: $A\mathbf{u}^{(1)} \neq \mathbf{b}$.

A better guess ought to be obtained by using the current guess and solving the first equation for a new $u_1^{(2)}$, the second for a new $u_2^{(2)}$, and so on. This gives us

$$\begin{aligned} 4u_1^{(2)} + 1 &= 1, \\ 2 + 5u_2^{(2)} + 1 &= 0, \\ -1 + 2 + 4u_3^{(2)} &= 3, \end{aligned}$$

and thus

$$\mathbf{u}^{(2)} = \begin{bmatrix} 0 \\ -0.6 \\ 0.5 \end{bmatrix}.$$

The next iteration becomes

$$\begin{aligned} 4u_1^{(3)} - 0.6 &= 1, \\ 5u_2^{(3)} + 0.5 &= 0, \\ -1.2 + 4u_3^{(3)} &= 3, \end{aligned}$$

and thus

$$\mathbf{u}^{(3)} = \begin{bmatrix} 0.4 \\ -0.1 \\ 1.05 \end{bmatrix}.$$

After a few more iterations, we will be close enough to the true solution

$$\mathbf{u} = \begin{bmatrix} 0.333 \\ -0.333 \\ 1.0 \end{bmatrix}.$$

Try one more iteration for yourself.

This iterative method is known as the *Gauss-Jacobi iteration.* In the general case, we are given a linear system with n equations and n unknowns u_i; $i = 1, \ldots, n$, written in matrix form as

$$A\mathbf{u} = \mathbf{b}.$$

Let us also assume that we have an initial guess $\mathbf{u}^{(1)}$ for the solution vector $\mathbf{u}$.

We now define two matrices D and R as follows: D is the diagonal matrix whose diagonal elements are those of A, and R is the matrix obtained from A by setting all its diagonal elements to zero. Clearly then,

$$A = D + R$$

and our linear system becomes

$$D\mathbf{u} + R\mathbf{u} = \mathbf{b}$$

or

$$\mathbf{u} = D^{-1}[\mathbf{b} - R\mathbf{u}].$$

In the spirit of our previous development, we now write this as

$$\mathbf{u}^{(k+1)} = D^{-1}[\mathbf{b} - R\mathbf{u}^{(k)}],$$

meaning that we attempt to compute a new estimate $\mathbf{u}^{(k+1)}$ from an existing one $\mathbf{u}^{(k)}$. Note that D must not contain zeroes on the diagonal; this can be achieved by row or column interchanges.

With this new framework, let us reconsider our last example. We have

$$A = \begin{bmatrix} 4 & 1 & 0 \\ 2 & 5 & 1 \\ -1 & 2 & 4 \end{bmatrix}, \quad R = \begin{bmatrix} 0 & 1 & 0 \\ 2 & 0 & 1 \\ -1 & 2 & 0 \end{bmatrix}, \quad D^{-1} = \begin{bmatrix} 0.25 & 0 & 0 \\ 0 & 0.2 & 0 \\ 0 & 0 & 0.25 \end{bmatrix}.$$

Then

$$\mathbf{u}^{(2)} = \begin{bmatrix} 0.25 & 0 & 0 \\ 0 & 0.2 & 0 \\ 0 & 0 & 0.25 \end{bmatrix} \left(\begin{bmatrix} 1 \\ 0 \\ 3 \end{bmatrix} - \begin{bmatrix} 0 & 1 & 0 \\ 2 & 0 & 1 \\ -1 & 2 & 0 \end{bmatrix} \begin{bmatrix} 1 \\ 1 \\ 1 \end{bmatrix} \right) = \begin{bmatrix} 0 \\ -0.6 \\ 0.5 \end{bmatrix}.$$

Will the Gauss-Jacobi method succeed? This is, will the sequence of vectors $\mathbf{u}^{(k)}$ converge? The answer is sometimes yes and sometimes no. It will *always* succeed if A is *diagonally dominant*,[3] and then it will succeed no matter what our initial guess $\mathbf{u}^{(1)}$ was. Many practical problems result in diagonally dominant systems.

In a practical setting, how do we determine if convergence is taking place? Ideally, we would like $\mathbf{u}^{(k)} = \mathbf{u}$, the true solution, after a number of iterations. Equality will most likely not happen, but the length of the *residual vector*

$$\|A\mathbf{u}^{(k)} - \mathbf{b}\|$$

should become small (i.e., less than some preset tolerance). Thus we check the size of the residual vector after each iteration, and stop once it is smaller than some preset tolerance.

A modification of the Gauss-Jacobi method is known as *Gauss-Seidel* iteration. When we compute $\mathbf{u}^{(k+1)}$ in the Gauss-Jacobi method, we can observe the following: the second element, $u_2^{(k+1)}$, is computed by using $u_1^{(k)}, u_3^{(k)}, \ldots, u_n^{(k)}$. We had just computed $u_1^{(k+1)}$. It stands to reason that using it instead of $u_1^{(k)}$ would be advantageous. This idea gives rise to the Gauss-Seidel method: as soon as a new element $u_i^{(k+1)}$ is computed, the estimate vector $\mathbf{u}^{(k+1)}$ is updated.

In summary, the Gauss-Jacobi method updates the new estimate vector once all of its elements are computed and Gauss-Seidel updates as soon as a new element is computed. Typically, Gauss-Seidel iteration converges faster than Gauss-Jacobi iteration.

5.7 Case Study: Fluid Flow

Before a submarine is built, its design is tested by a battery of computer simulations. These include stability tests, depth range tests, fuel efficiency tests, among others. Many of these simulations—from

[3]A matrix is diagonally dominant if for every row, the absolute value of its diagonal element is larger than the sum of the absolute values of its remaining elements.

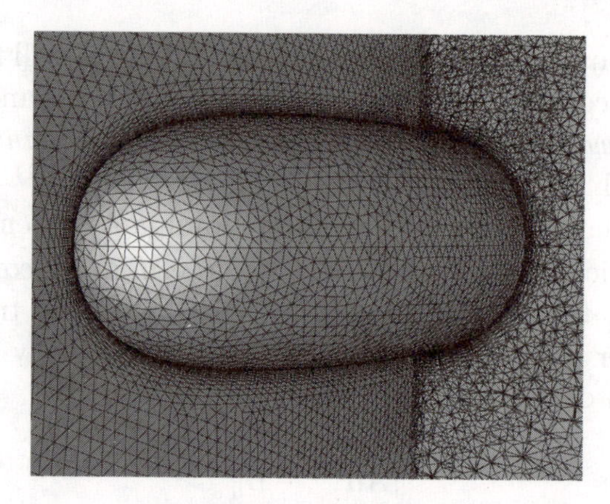

Figure 5.2. Grids for use in submarine fluid flow computation. (Courtesy of R. Loehner, George Mason University.)

the field of *computational fluid dynamics* (CFD)—employ finite element methods for solving the corresponding partial differential equations (PDEs). Figure 5.2 shows three meshes; each is a triangle mesh (see Section 13.5 for more on triangle meshes).

Let $\mathbf{x}_i$ be a vertex in the mesh, and let $\mathbf{y}_1, \ldots, \mathbf{y}_N$ denote its N neighbors (the points connected to $\mathbf{x}_i$). Modeling flow around the submarine quickly leads to solving PDEs. Then equations of the form

$$\mathbf{x}_i = \alpha_1 \mathbf{y}_1 + \ldots + \alpha_N \mathbf{y}_N$$

arise from discretizations[4] of those PDEs. For each vertex $\mathbf{x}_i$, we will thus have one equation, each involving six neighbors (on average). Given that the grid typically has several thousand vertices, this means that each row of the resulting matrix has about six nonzero entries and several thousand zero entries. Such matrices are called *sparse*.

The sparse matrix M is stored by using a pointer system. If matrix entry $m_{i,j}$ is nonzero, then this value will be stored together with the integer pair i, j. Assuming around ten nonzero entries per row, and a size of $10,000 \times 10,000$ for M, we have to store $10 \times 10,000 = 10^5$ reals and 2×10^5 integers. Compare to storing the whole array with 10^8 reals!

[4]A discretization replaces a continuous model by one that is broken down into many (discrete) pieces.

5.8 Overdetermined Systems

If we want to determine the values of n parameters from some linear model, then typically we try to obtain n data in order to solve a linear system. Sometimes there are more data than unknowns—this leads to overdetermined systems.

An overdetermined linear system has n equations in m unknowns, where $m > n$. It looks like this:

$$\begin{bmatrix} a_{1,1} & \cdots & a_{1,n} \\ \vdots & & \vdots \\ \vdots & & \vdots \\ \vdots & & \vdots \\ a_{m,1} & \cdots & a_{m,n} \end{bmatrix} \begin{bmatrix} u_1 \\ \vdots \\ u_n \end{bmatrix} = \begin{bmatrix} b_1 \\ \vdots \\ \vdots \\ \vdots \\ b_m \end{bmatrix}. \tag{5.4}$$

This also has the matrix form

$$A\mathbf{u} = \mathbf{b}. \tag{5.5}$$

However, A is not square. An overdetermined system, as the name implies, will in general not have a solution. Yet an approximate solution may be found as follows. If we multiply both sides of (5.5) by A's transpose, A^{T}, we obtain

$$A^{\mathrm{T}}A\mathbf{u} = A^{\mathrm{T}}\mathbf{b}. \tag{5.6}$$

The matrix $A^{\mathrm{T}}A$ is square and typically also nonsingular. The linear system (5.6) is known as the *normal equations*, which form a standard linear system that may be solved by Gauss elimination. The exact solution $\mathbf{u}$ of this system is only an approximate solution to the original system (5.5). It can be shown, though, that it is in fact the best possible one with respect to the L^2 norm; that is, $||A\mathbf{u}-\mathbf{b}||^2$ will be minimized, and this is called the *least squares solution.* We revisit this topic in several sections of Chapter 8 on data fitting. A different solution strategy is via SVD, as discussed in Section 6.8.

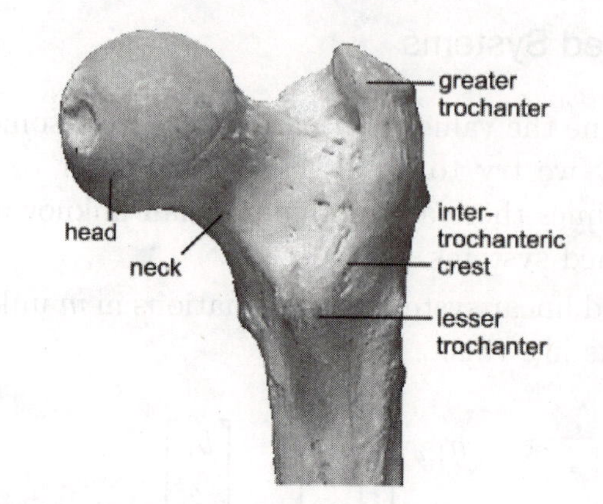

Figure 5.3. A femoral head. (Courtesy of Biological and Life Sciences, University of Glascow.)

5.9 Case Study: Femoral Head Reconstruction

Physical anthropologists studying human evolution have to work with bones or bone fragments and determine their origin and functionality. To determine whether parts of a skeleton belonged to a bipedal specimen, calls for careful analysis of the pelvic bones, among them the exact shape of the femoral head, which is the spherical bone on top of the femur, as shown in Figure 5.3.

When working with bones, it is imperative that modern anthropologists know the shape of the bones in an exact way. To this end, they employ *digitizers* which produce a series of coordinates of points on a bone.

In the case of a femur, an important question is how spherical it is. Thus, let us say we have digitized[5] n coordinate triples $\mathbf{x}_i$ and we want to find a best-fitting sphere. That sphere would have an implicit equation of the form

$$x^2 + y^2 + z^2 + bx + cy + dz + e = 0.$$

Ideally there are parameters b, c, d, e such that all $\mathbf{x}_i$ satisfy this equation. Then we would have the system

[5]With mechanical digitizers, n would range in the low hundreds, and for laser digitizers, in the thousands.

$$
\begin{aligned}
x_1^2 + y_1^2 + z_1^2 + bx_1 + cy_1 + dz_1 + e &= 0\\
&\vdots\\
&\vdots\\
x_n^2 + y_n^2 + z_n^2 + bx_n + cy_n + dz_n + e &= 0.
\end{aligned}
$$

The squared terms are nothing to worry about—they are all known. In fact, if we move those terms over to the right-hand side, get

$$
\begin{aligned}
bx_1 + cy_1 + dz_1 + e &= -x_1^2 - y_1^2 - z_1^2\\
&\vdots\\
&\vdots\\
bx_n + cy_n + dz_n + e &= -x_n^2 - y_n^2 - z_n^2.
\end{aligned}
$$

In matrix form:

$$
\begin{bmatrix} x_1 & y_1 & z_1 & 1\\ \vdots & & & \vdots\\ \vdots & & & \vdots\\ x_n & y_n & z_n & 1 \end{bmatrix}
\begin{bmatrix} b\\ c\\ d\\ e \end{bmatrix}
=
\begin{bmatrix} -x_1^2 - y_1^2 - z_1^2\\ \vdots\\ \vdots\\ -x_n^2 - y_n^2 - z_n^2 \end{bmatrix}
$$

or, even shorter:

$$X\mathbf{a} = \mathbf{s}.$$

We have an overdetermined linear system with n equations for the unknowns b, c, d, e. It is readily solved by using the normal equation approach.

The parameters b, c, d, e fix a sphere. It is now trivial to compute the sphere's center $\mathbf{c}$ and radius r. Then we can easily determine how well the $\mathbf{x}_i$ fit the sphere by inspecting the deviations $\|\mathbf{x}_i - \mathbf{c}\|^2 - r^2$.

5.10 Problems and Experiments

1. Design an unstable 2×2 linear system in the sense of Section 5.4.

2. An $n \times n$ *Hilbert matrix* H_n is defined by its elements $h_{i,j}$,

$$h_{i,j} = \frac{1}{i+j-1}, \quad 1 \le i,j \le n.$$

Using your favorite package, compute the determinants $|H_n|$ for $n = 1, 20$. (Based upon those numbers, you should be worried about the idea of having to work with these matrices!)

3. A plane may be expressed as

$$ax + by + cz = d.$$

Suppose you are given three planes $a_i x + b_i y + c_i z = d_i; i = 1, 2, 3$. Finding the intersection point of these planes leads to solving a linear system. Discuss under what condition there is no solution to this system. Some hints may be found in Section 13.3.3.

4. Let $a_i x = b_i; i = 1, \ldots, 100$ be an overdetermined linear system for just one unknown x. Find an explicit solution for x.

6

Eigen-Problems

This chapter introduces the world of eigenvalues and eigenvectors. In German, the term *eigen* means characteristic or special. As we see in this chapter, eigenvalues and eigenvectors reveal key characteristics of the action of a square matrix. Eigen-analysis is very important in science and engineering. For example, eigenvalues are related to frequencies, and as an application, the stability of a bridge can be predicted by an eigenvalue analysis of a system modeling it. Figure 6.1 illustrates the collapse of the Tacoma Narrows bridge in 1940 due to wind-induced vibrations.

Several methods for computing solutions to eigen-problems are introduced. The power method is a means for computing the largest

Figure 6.1. The collapse of the Tacoma Narrows bridge. (Courtesy of Doug Smith, Carleton University.)

eigenvalue and its corresponding eigenvector. Jacobi iteration is a means for computing all eigenvalues and eigenvectors. For nonsquare matrices, a generalization of eigenvalues are singular values. The singular value decomposition is a useful tool for analyzing matrices and for solving linear systems. Many applications utilize the singular value decomposition as part of a principal components analysis. This type of analysis reduces a problem size by distilling a large set of vectors down to a smaller, manageable set.

6.1 Eigenvalues

On the left of Figure 6.2 a star image appears with arrows; the ones parallel to the coordinate axes are highlighted in gray. We then applied the matrix

$$A = \begin{bmatrix} 2 & 0.5 \\ 0 & 1 \end{bmatrix}$$

to all arrows; the result is on the right of Figure 6.2.

Visual inspection shows that A maps horizontal vectors to horizontal vectors, that is, to multiples of themselves. A matrix A having a special vector $\mathbf{u}$ such that

$$A\mathbf{u} = \lambda\mathbf{u} \tag{6.1}$$

is said to have an *eigenvector* $\mathbf{u}$. The scaling factor λ is called A's *eigenvalue* for $\mathbf{u}$. Matrices of size $n \times n$ have up to n linearly inde-

Figure 6.2. All arrows on the left are mapped by a shear to the arrows on the right. Only the two horizontal arrows are mapped to multiples of themselves.

pendent eigenvectors. Not all matrices have eigenvectors however; for example, a 2D rotation does not map any vector to itself nor to a multiple of itself.

Eigenvalues are at the core of computing with matrices. In most cases, scientists are confronted with eigenvalue problems for *symmetric matrices*[1] having the property $A^{\mathrm{T}} = A$. For these matrices, all eigenvalues λ_i are real; for nonsymmetric matrices, eigenvalues may be complex numbers (this will become clear in Section 6.5). Furthermore, all eigenvectors $\mathbf{u}_i$ of a symmetric matrix are orthogonal. If we normalize all $\mathbf{u}_i$ and consider them as column vectors in a matrix U, then U will be *orthonormal*: $U^{-1} = U^{\mathrm{T}}$.

For example, let

$$A = \begin{bmatrix} 2 & 1 \\ 1 & 2 \end{bmatrix}.$$

Using a package such as Mathematica, we find two eigenvalues, $\lambda_1 = 3$ and $\lambda_2 = 1$, and eigenvectors $\mathbf{u}_1 = [1, -1]^{\mathrm{T}}$ and $\mathbf{u}_2 = [1, 1]^{\mathrm{T}}$. We form U by normalizing the eigenvectors and making them the column vectors:

$$U = \begin{bmatrix} 0.707 & 0.707 \\ -0.707 & 0.707 \end{bmatrix}.$$

Defining a matrix Λ as[2]

$$\Lambda = \begin{bmatrix} 3 & 0 \\ 0 & 1 \end{bmatrix},$$

we see that

$$AU = U\Lambda, \tag{6.2}$$

capturing all eigenvalue equations in one matrix equation. Equation (6.2) holds not only for our example: it is the defining equation for all eigenvalues and eigenvectors of symmetric matrices for any dimension n. As a consequence, we also have

$$A = U\Lambda U^{\mathrm{T}}. \tag{6.3}$$

Here, you have to recall that $U^{-1} = U^{\mathrm{T}}$ for orthonormal matrices.

[1]Symmetric matrices are formed in least squares problems, for instance. In these situations, we are given more data than needed for a solution.

[2]Λ is the capital Greek letter "Lambda."

6.2 The Power Method

The definition of eigenvalues is easy—but how does one compute them? In many cases, only the eigenvalue λ_1 with the largest absolute value (called the *dominant eigenvalue*) is desired (see Section 6.3). The corresponding eigenvector is called the *dominant eigenvector*. Both, the dominant eigenvalue and the dominant eigenvector, are found by using the *power method*.

Starting with an example, we reuse the matrix

$$A = \begin{bmatrix} 2 & 1 \\ 1 & 2 \end{bmatrix}.$$

Let's apply it repeatedly to the vector $\mathbf{v}^{(1)} = [1, 2]^{\mathrm{T}}$. We obtain the vector sequence $\mathbf{v}^{(i)} = A^i\mathbf{v}$:

$$\begin{bmatrix} 1 \\ 2 \end{bmatrix}, \begin{bmatrix} 4 \\ 5 \end{bmatrix}, \begin{bmatrix} 13 \\ 14 \end{bmatrix}, \ldots, \begin{bmatrix} 29524 \\ 29525 \end{bmatrix}, \begin{bmatrix} 88573 \\ 88574 \end{bmatrix}, \ldots.$$

If we take ratios of successive x-components, we obtain $4/1, 13/4, \ldots, 88573/29524, \ldots$, or in terms of real numbers: $4, 3.25, \ldots, 3.0003, \ldots$. Repeating for the y-components gives $2.5, 2.8, \ldots, 2.99997, \ldots$. Thus, for sufficiently large i, we have $\mathbf{v}^{(i)} = 3\mathbf{v}^{(i-1)}$. You might recall that A's largest eigenvalue is 3.

To understand this better, recall that a symmetric matrix has n orthonormal eigenvectors $\mathbf{u}_1, \ldots, \mathbf{u}_n$ and corresponding eigenvalues $\lambda_1, \ldots \lambda_n$. The $\mathbf{u}_i$ form a basis for $\mathbb{R}^n$, and any vector may be expressed in terms of it. Let us express the vector $\mathbf{v}^{(1)}$ in terms of the $\mathbf{u}_i$:

$$\mathbf{v}^{(1)} = c_1\mathbf{u}_1 + c_2\mathbf{u}_2 + \ldots + c_n\mathbf{u}_n.$$

The vector $\mathbf{v}^{(2)}$ then is

$$\mathbf{v}^{(2)} = A\mathbf{v}^{(1)} = c_1\lambda_1\mathbf{u}_1 + c_2\lambda_2\mathbf{u}_2 + \ldots + c_n\lambda_n\mathbf{u}_n.$$

If we repeat the process:

$$\mathbf{v}^{(i)} = A\mathbf{v}^{(i-1)} = c_1\lambda_1^{i-1}\mathbf{u}_1 + c_2\lambda_2^{i-1}\mathbf{u}_2 + \ldots + c_n\lambda_n^{i-1}\mathbf{u}_n.$$

Since we assumed that λ_1 is the largest eigenvalue (in absolute value), the term $c_1\lambda_1^{i-1}\mathbf{u}_1$ will dominate the expression for $\mathbf{v}^{(i)}$, thus essentially making $\mathbf{v}^{(i)}$ parallel to $\mathbf{u}_1$.

Given: A symmetric $n \times n$ matrix A.
Wanted: A's largest eigenvalue (in absolute value).

Step 1: Pick an arbitrary nonzero n-vector $\mathbf{v}^{(0)}$.

Step 2: Create a sequence $\mathbf{v}^{(i)} = A\mathbf{v}^{(i-1)}$.

Step 3: Once the ratios of analogous components converge to a limit λ_1, that limit is the desired eigenvalue.

Step 4: The final vector of the sequence should be normalized and then this is the eigenvector corresponding to λ_1.

Here, then, is the power method algorithm:
Figure 6.3 illustrates the sequence of normalized 2D vectors of our example.

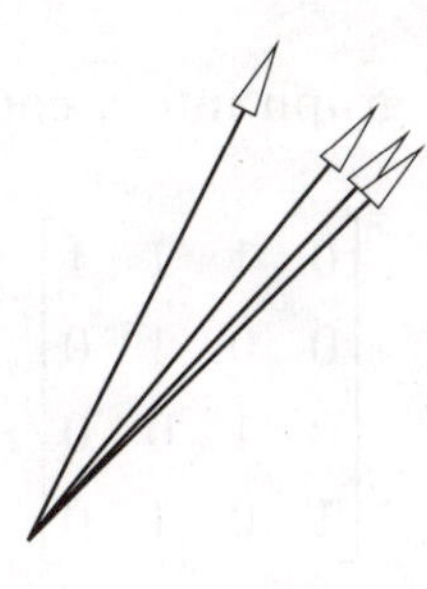

Figure 6.3. A converging sequence of unit vectors.

The speed of convergence depends on the size of the second largest eigenvalue λ_2. If λ_2 were identical to λ_1, the above convergence argument would not work. If both eigenvalues were close, convergence would be slow. Thus convergence is a function of $|\lambda_1|/|\lambda_2|$: the larger this ratio, the speedier the convergence process.

6.3 Case Study: PageRank

The World Wide Web, at some frozen point in time, consists of N webpages, many of them pointing to (having links to) other web-

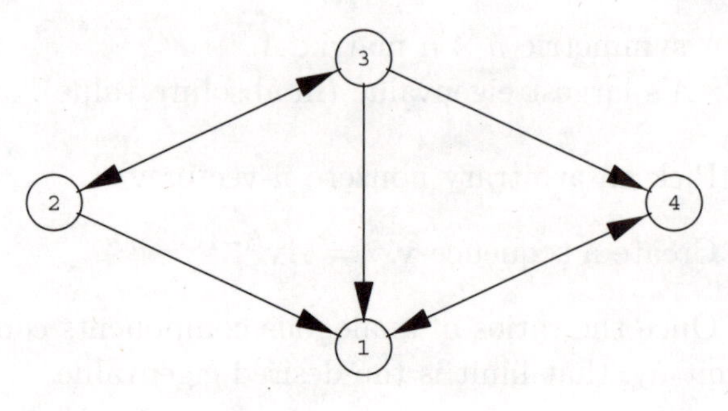

Figure 6.4. A micro-web.

pages.[3] A page that is pointed to very often would be considered important, whereas a page with very few other pages pointing to it would be considered not important. How can we rank all webpages according to how important they are? In the sequel, we assume that all webpages are ordered in some fashion, such as lexicographic, so we can assign a number i to every page. An example of a "micro-web" is shown in Figure 6.4.

We may transcribe this graph into a connectivity matrix C:

$$C = \begin{bmatrix} 0 & 1 & 1 & 1 \\ 0 & 0 & 1 & 0 \\ 0 & 1 & 0 & 0 \\ 1 & 0 & 1 & 0 \end{bmatrix}.$$

In this example, page 3 links to page 1, hence $c_{1,3} = 1$. Page 2 does not link to page 4, hence $c_{4,2} = 0$.

The connectivity matrix is a concept that has been known since before the World Wide Web. However, it was Google that for the first time exploited its underlying structure to create PageRank, the page ranking system that is the root of Google's claim to fame.

Let l_j be the total number of pages that page j links to. This is the sum of all elements of the jth column of C. The more links page j has, the lower will be its contribution to those pages. Thus, we scale every element of column j by $1/l_j$. The resulting matrix D

[3] Currently, $N = 30$ billion.

is given by

$$d_{i,j} = \frac{c_{i,j}}{l_j}.$$

Note that all columns of D have nonnegative entries and sum to one. Matrices with that property are called *stochastic.*

In the example above, we get

$$D = \begin{bmatrix} 0 & 1/2 & 1/3 & 1 \\ 0 & 0 & 1/3 & 0 \\ 0 & 1/2 & 0 & 0 \\ 1 & 0 & 1/3 & 0 \end{bmatrix}.$$

The ranking r_j of page j thus depends on all pages linking to it. Let's call the set of those pages L_j.

$$r_j = \sum_{i \in L_j} \frac{r_i}{l_i}.$$

Using the vector $\mathbf{r} = [r_1, \ldots, r_N]^{\mathrm{T}}$, we get

$$\mathbf{r} = D\mathbf{r}. \tag{6.4}$$

This states that we are looking for the eigenvector of D corresponding to the eigenvalue 1. All stochastic matrices have 1 as their largest eigenvalue. In order to find $\mathbf{r}$, we employ the power method from Section 6.2. This method needs an initial guess for $\mathbf{r}$, and setting all $r_i = 1$ is not too bad for that. As the iterations converge, the solution is found.

The vector $\mathbf{r}$ now contains the ranking—called PageRank by Google—of every page. In our example, $\mathbf{r} = [0.71, 0, 0, 0.71]^{\mathrm{T}}$, thus pages 1 and 4 are ranked highest.

The actual Google computation is more involved than our outline, but it already shows how the page rank problem can be attacked by using the power method.

6.4 Jacobi Iteration

The power method was useful for finding the largest eigenvalue and the corresponding eigenvector. There are problems, however, that require *all* eigenvalues and eigenvectors; an example is the singular value decomposition, addressed in Section 6.6. Here, we describe the Jacobi method, which iteratively transforms the matrix such that the diagonal elements converge to its eigenvalues and the off-diagonal elements converge to zero.

Let us start with a 2×2 matrix A. From (6.2), we know that A's eigenvalues reside in the matrix Λ, which is given by

$$\Lambda = U^{\mathrm{T}} A U.$$

U is a rotation matrix and U^{T} is the inverse rotation. If we knew the rotation angle α, then we could simply build U from

$$\begin{bmatrix} \cos\alpha & -\sin\alpha \\ \sin\alpha & \cos\alpha \end{bmatrix}.$$

It turns out that α can be computed from

$$\tan 2\alpha = \frac{2a_{1,2}}{a_{1,1} - a_{2,2}}.$$

Let's try this for our matrix

$$A = \begin{bmatrix} 2 & 1 \\ 1 & 2 \end{bmatrix}.$$

We find $\tan 2\alpha = 4/0 = \infty$; hence $\alpha = 45°$. Then

$$U = \begin{bmatrix} 0.707 & -0.707 \\ 0.707 & -0.707 \end{bmatrix}.$$

We compute

$$U^{\mathrm{T}} A U = \begin{bmatrix} 2.999 & 0 \\ 0 & 0.999 \end{bmatrix},$$

which tells us the eigenvalues are 2.999 and 0.999. Since we used only four digits, this is close enough to the correct values 3 and 1.

This method can be applied to a larger symmetric matrix A as follows. We first note that a matrix of the form

$$U = \begin{bmatrix} 1 & & & & & & \\ & \ddots & & & & & \\ & & \cos\alpha & \cdots & -\sin\alpha & & \\ & & \vdots & \ddots & \vdots & & \\ & & \sin\alpha & \cdots & \cos\alpha & & \\ & & & & & \ddots & \\ & & & & & & 1 \end{bmatrix}$$

is orthonormal. (Check for a small example!) Furthermore, if $u_{i,j} = -\sin\alpha$, then the matrix $A' = U^{\mathrm{T}}AU$ has zero elements $a'_{i,j} = a_{j,i} = 0$. The idea now is to successively zero out elements of A. Unfortunately, elements that had been zero may be nonzero in the next step. However, the algorithm is guaranteed to converge.

We summarize by stating the Jacobi iteration method.

Given: a symmetric matrix $A^{(1)}$. *Wanted*: its eigenvalues $\lambda_1, \ldots, \lambda_n$.

Step 1: Pick the largest off-diagonal element $a_{i,j}$.

Step 2: Compute an angle α from $\tan 2\alpha = \dfrac{2a_{i,j}}{a_{i,i} - a_{j,j}}$ and form the matrix $U^{(1)}$.

Step 3: Set $A^{(2)} = U^{(1)\mathrm{T}} A^{(1)} U^{(1)}$.

Step 4: Repeat, creating a sequence $A^{(r)}$, and stop once all off-diagonal elements are below a given threshold, say $r = N$.

Then $A^{(N)}$ has $A^{(1)}$'s eigenvalues on its diagonal and the product matrix $U^{(1)} \cdots U^{(N)}$ holds $A^{(1)}$'s eigenvectors.

For example, we let

$$A^{(1)} = \begin{bmatrix} 3 & -1 & 1 \\ -1 & 4 & 2 \\ 1 & 2 & -3 \end{bmatrix}.$$

We select to zero out $a_{2,3}^{(1)}$:

$$A^{(2)} = \begin{bmatrix} 3 & -0.709 & 1.22 \\ -0.709 & 4.53 & 0 \\ 1.22 & 0 & -3.53 \end{bmatrix}.$$

Note the two zeroes in the right places! Next:

$$A^{(3)} = \begin{bmatrix} 3.22 & -0.698 & -0. \\ -0.698 & -4.5 & -0.126 \\ -0. & -0.126 & -3.75 \end{bmatrix}.$$

Continuing:

$$A^{(4)} = \begin{bmatrix} 2.92 & 0. & 0.05 \\ 0. & 4.83 & 0.116 \\ 0.05 & 0.116 & -3.75 \end{bmatrix}.$$

We could continue, but we are already very close to the correct eigenvalues of $4.835, -3.755, 2.919$.

6.5 Eigenvalues and Determinants

In classical linear algebra books, eigenvalues are typically introduced as follows. The defining equation, using an $n \times n$ matrix A, is

$$A\mathbf{v} = \lambda\mathbf{v}$$

where $\mathbf{v}$ is not the zero vector and λ is a nonzero number. We immediately obtain

$$[A - \lambda I]\mathbf{v} = \mathbf{0}.$$

Thus the matrix $A - \lambda I$ maps a nonzero vector $\mathbf{v}$ to the zero vector $\mathbf{0}$; it therefore must be singular. Singular matrices have zero determinants:

$$|A - \lambda I| = 0.$$

Clearly, $|A - \lambda I|$ is a function of the unknown λ. In fact it is a polynomial of degree n with the variable λ. Thus, we have reduced the eigenvalue problem to finding the zeroes of a polynomial. If A is symmetric, these zeroes will be real; otherwise, they may be complex numbers. If A is $n \times n$ and n is odd, then we are guaranteed to have at least one real eigenvalue because any odd-degree polynomial has at least one real root.

A quick example reuses A from Section 6.4:

$$A = \begin{bmatrix} 3 & -1 & 1 \\ -1 & 4 & 2 \\ 1 & 2 & -3 \end{bmatrix}.$$

Then

$$|A - \lambda I| = \begin{vmatrix} 3-\lambda & -1 & 1 \\ -1 & 4-\lambda & 2 \\ 1 & 2 & -3-\lambda \end{vmatrix} = -\lambda^3 + 2\lambda^2 + 17\lambda - 21.$$

This cubic polynomial in λ has the three zeroes, 4.835, −3.755, 2.919, which we already had found as A's eigenvalues.

Although this approach is theoretically feasible, we have traded a manageable problem (eigenvalues) for a much more difficult one (zeroes of polynomials). The zeroes of low degree polynomials pose no serious threat, but matrices of size $1,000 \times 1,000$ would require finding the zeroes of a degree 1,000 polynomial—an impossible task numerically. See Section 10.5 for an example.

6.6 Singular Value Decomposition

One of the central methods in applied numerical linear algebra is singular value decomposition. A surprisingly simple question will be the key to our investigation: What set of orthogonal vectors is taken to another set of orthogonal vectors?

Let A be a rectangular matrix with m rows and n columns. This means A maps vectors from $\mathbb{R}^n$ to vectors in $\mathbb{R}^m$. Let $\mathbf{u}_1, \ldots, \mathbf{u}_n$ be a set of orthonormal vectors in A's domain $\mathbb{R}^n$, meaning that the matrix $U = [\mathbf{u}_1, \ldots, \mathbf{u}_n]$ is orthonormal: $U^{-1} = U^{\mathrm{T}}$ (see Section 4.4). In general, these vectors will not be mapped to n orthogonal vectors $\mathbf{w}_1, \ldots, \mathbf{w}_n$ in A's range $\mathbb{R}^m$.[4] As it turns out, a special set of $\mathbf{u}_i$ *will*

[4]Note that the $\mathbf{u}_i$ each have n elements, whereas the $\mathbf{w}_i$ each have m elements.

be mapped to a set of orthogonal vectors $\mathbf{w}_i$, forming an orthogonal matrix W. We express W as the product of an orthonormal matrix V and a diagonal matrix Σ, that is, $W = V\Sigma$.

We thus have

$$AU = V\Sigma, \tag{6.5}$$

where Σ is a (generalized) diagonal matrix with m rows and n columns, having only nonzero elements, σ_i, on its diagonal. V is $m \times m$, and recall that U is $n \times n$. Using (6.5), we obtain

$$A = V\Sigma U^{-1} = V\Sigma U^{\mathrm{T}}. \tag{6.6}$$

This is the *singular value decomposition* (SVD) of the rectangular matrix A. The σ_i are the *singular values* of A. The matrix dimensions are illustrated in Figure 6.5.

Before showing how to obtain U, Σ, V, let's look at an example. Let A and $A^{\mathrm{T}}A$ be given by

$$A = \begin{bmatrix} 1 & 0 \\ 0 & 2 \\ 0 & 1 \end{bmatrix}, \quad A^{\mathrm{T}}A = \begin{bmatrix} 1 & 0 \\ 0 & 5 \end{bmatrix}. \tag{6.7}$$

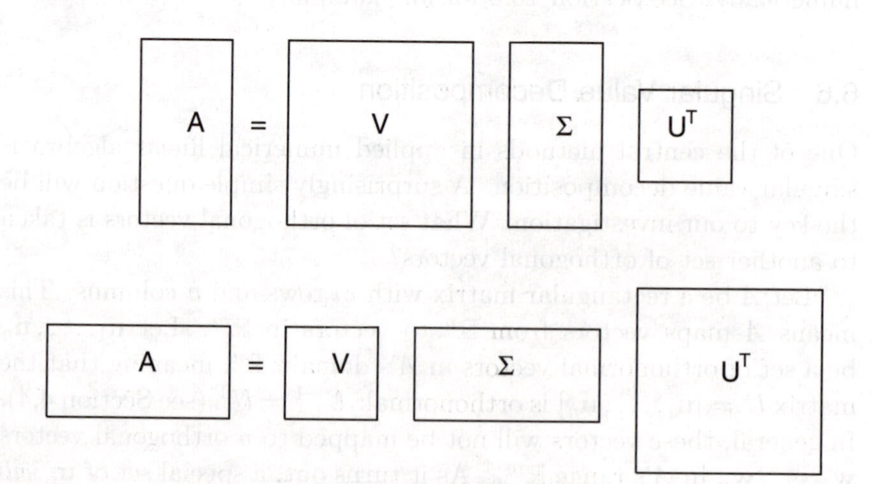

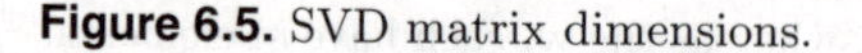

Figure 6.5. SVD matrix dimensions.

Then (using a package such as Mathematica) we find

$$\Sigma = \begin{bmatrix} \sqrt{5} & 0 \\ 0 & 1 \\ 0 & 0 \end{bmatrix}, \quad U = \begin{bmatrix} 0 & 1 \\ 1 & 0 \end{bmatrix}, \quad V = \begin{bmatrix} 0 & 1 & 0 \\ \frac{2}{\sqrt{5}} & 0 & -\frac{1}{\sqrt{5}} \\ \frac{1}{\sqrt{5}} & 0 & \frac{2}{\sqrt{5}} \end{bmatrix}.$$

It is now trivial to verify $AU = V\Sigma$.

So, how do we find U, Σ, V? We multiply both sides of (6.5) by their transposes:

$$(AU)^{\mathrm{T}} AU = (V\Sigma)^{\mathrm{T}} V\Sigma.$$

Hence,

$$U^{\mathrm{T}} A^{\mathrm{T}} AU = \Sigma^{\mathrm{T}} V^{\mathrm{T}} V\Sigma,$$

and finally,

$$A^{\mathrm{T}} A = U\Lambda U^{\mathrm{T}}, \tag{6.8}$$

where Λ is the $m \times m$ matrix $\Lambda = \Sigma^{\mathrm{T}}\Sigma$. Equation (6.8) states the following: The symmetric matrix $A^{\mathrm{T}}A$ has eigenvalues, λ_i, which are the diagonal entries of Λ, and eigenvectors, which are the columns of U; see (6.3). Jacobi iteration from Section 6.4 gives us the means to compute U and Λ. The σ_i are given by $\sigma_i = \sqrt{\lambda_i}$, from which we can build Σ. Finally, $V = AU\Sigma^+$ where Σ^+ denotes the pseudoinverse of the generalized diagonal matrix Σ; see Section 6.8.

If $A^{\mathrm{T}}A$ has full rank, then all the σ_i are strictly positive. One typically arranges the singular values such that $\sigma_1 \geq \sigma_2 \geq \cdots \geq \sigma_n$.

The ratio σ_1/σ_n (assuming $\sigma_1 \geq \sigma_n$) is called the *condition number* of A. The larger the condition number, the more A distorts. If the condition number is 1, such as for the identity matrix, then no distortion occurs. As an exercise, find out why 1 is the lowest possible condition number for a nonsingular matrix.

To see why this makes sense, let's look at two examples. First, let

$$A = \begin{bmatrix} \cos\alpha & -\sin\alpha \\ \sin\alpha & \cos\alpha \end{bmatrix},$$

meaning that A is a rotation about α degrees. Clearly, $A^{\mathrm{T}}A = I$, where I is the identity matrix. Thus, $\sigma_1 = \sigma_2 = 1$. Hence the condition of a rotation matrix is 1. Since a rotation does not distort at all, this is quite intuitive.

Now let

$$A = \begin{bmatrix} 100 & 0 \\ 0 & 0.01 \end{bmatrix},$$

a matrix that scales by 100 in the x-direction and by 0.01 in the y-direction. This matrix is severely distorting! We quickly find $\sigma_1 = 100, \sigma_2 = 0.01$, and thus the condition number is $100/0.01 = 10,000$. The fact that A distorts is clearly mirrored by the magnitude of its condition number. The topic of condition numbers is revisited in Section 6.7.

6.7 The Condition Number

In Section 6.6, the condition number of a matrix was defined as the ratio of the largest to smallest singular values. Some matrices, such as orthonormal ones, are "well-behaved," whereas others, close to being singular, are known to cause numerical problems. A singular matrix has at least one singular value of 0, resulting in a ratio of the largest to smallest singular values of ∞. For an orthonormal matrix, that ratio is 1. It stands to reason that a high ratio of largest to smallest singular values is "bad."

Let us now revisit our linear system from Section 5.4. We saw that a small change in the right-hand side could lead to large changes in the solution. The coefficient matrix was

$$A = \begin{bmatrix} 0.8 & 0.1 & 0.1 \\ 0.7 & 0.24 & 0.06 \\ 0.6 & 0.4 & 0 \end{bmatrix}. \tag{6.9}$$

Its singular values are $1.3, 0.26, 0.006$. The largest and smallest singular values are very different in size. Their ratio is 215, a number indicating that the matrix is close to being singular. Hence the unstable behavior!

6.8 The Pseudoinverse

The inverse of a matrix A is another matrix A^{-1} such that $AA^{-1} = I$, the identity matrix. Not every matrix A has an inverse; in fact, A

must be square and invertible. In this section, we explore how to generalize the concept of an inverse matrix.

Singular value decomposition will allow us to define an "inverse" of a nonsquare matrix. In preparation, let's consider inverses of diagonal matrices. A square diagonal matrix is inverted by computing the reciprocals of its nonzero elements; see (4.14). A nonsquare diagonal matrix D is not invertible, but a similar process yields the *pseudoinverse* designated by D^+. Specifically, we take the reciprocals of the transpose matrix. This results in $DD^+ = I$. For example:

$$D = \begin{bmatrix} 4 & 0 & 0 \\ 0 & 2 & 0 \end{bmatrix}, \qquad D^+ = \begin{bmatrix} 0.25 & 0 \\ 0 & 0.5 \\ 0 & 0 \end{bmatrix}, \qquad DD^+ \begin{bmatrix} 1 & 0 \\ 0 & 1 \end{bmatrix}.$$

If we repeat the inversion process, the initial matrix D is recovered:

$$D^{++} = D.$$

Check for the above example! For an invertible diagonal matrix D, we have $D^+ = D^{-1}$. If D contains very small[5] nonzero elements, then these need to be set to zero in order to avoid extremely large entries in D^+.

We are now ready to define a pseudoinverse for general matrices. This pseudoinverse will have the property that it equals the inverse for invertible matrices. Since any matrix A may be written as $A = U\Sigma V^{\mathrm{T}}$ by using SVD, we define

$$A^+ = V\Sigma^+ U^{\mathrm{T}}. \tag{6.10}$$

For example, take

$$A = \begin{bmatrix} 2 & 0 & 0 \\ 0 & 0 & -4 \end{bmatrix} = \begin{bmatrix} 0 & 1 \\ -1 & 0 \end{bmatrix} \begin{bmatrix} 4 & 0 & 0 \\ 0 & 2 & 0 \end{bmatrix} \begin{bmatrix} 0 & 0 & 1 \\ 1 & 0 & 0 \\ 0 & 1 & 0 \end{bmatrix}.$$

Hence,

$$A^+ = \begin{bmatrix} 0.5 & 0 \\ 0 & 0 \\ 0 & -0.25 \end{bmatrix}.$$

[5] "Small" is a function of the application at hand; without such knowledge, 10^{-6} is a decent guess.

The pseudoinverse may be used to approximate solutions for linear systems. If the system

$$A\mathbf{u} = \mathbf{b}$$

does not have an invertible coefficient matrix A, then

$$\mathbf{u} = A^{+}\mathbf{b}$$

is an optimal approximation to a solution.

For example, using the matrix A from Equation 6.7, we create a linear system as follows:

$$\begin{bmatrix} 1 & 0 \\ 0 & 2 \\ 0 & 1 \end{bmatrix} \begin{bmatrix} x_1 \\ x_2 \end{bmatrix} = \begin{bmatrix} 0 \\ 1 \\ 0 \end{bmatrix}.$$

An approximate solution is given by

$$\mathbf{x} = A^{+}\mathbf{b} = \begin{bmatrix} 1 & 0 & 0 \\ 0 & 2/5 & 1/5 \end{bmatrix} \begin{bmatrix} 0 \\ 1 \\ 0 \end{bmatrix} = \begin{bmatrix} 0 \\ 2/5 \end{bmatrix}.$$

Note that this approximate solution to the overdetermined linear system is identical to that of the *normal equations*

$$A^{\mathrm{T}}A\mathbf{x} = A^{\mathrm{T}}\mathbf{b}.$$

See Section 5.8 for more information on normal equations.

6.9 The Principal Components Analysis

We may rewrite the SVD from (6.6) in the dyadic form of (4.6):

$$A = \sum_{i=1}^{r} \sigma_i \mathbf{v}_i \mathbf{u}_i^{\mathrm{T}}, \tag{6.11}$$

where r is the rank of A and we have omitted zero terms (in case A was not of full rank). Also, assume that we have ordered the summation terms in (6.11) so that $\sigma_i > \sigma_{i+1}$. Recall that $\mathbf{v}_i\mathbf{u}_i^{\mathrm{T}}$ is *not* a scalar but rather an $m \times n$ matrix. (The $\mathbf{u}_i^{\mathrm{T}}$ term is the ith column of U written as a row vector.)

Let us now approximate A by a matrix $\hat{A}$:

$$\hat{A} = \sum_{i=1}^{k} \sigma_i \mathbf{v}_i \mathbf{u}_i^{\mathrm{T}}, \tag{6.12}$$

where $k < r$. If k is much smaller than r, then we have approximated A by a matrix $\hat{A}$, which is defined by a much smaller number of data, namely the vectors $\mathbf{u}_1, \ldots, \mathbf{u}_k$; the vectors $\mathbf{v}_1, \ldots, \mathbf{v}_k$; and the singular values $\sigma_1, \ldots, \sigma_k$. This approximation makes sense only if the singular values $\sigma_{k+1}, \ldots, \sigma_r$ are small; however, this is the case in many applications. This reduction in dimensionality in (6.12) is called the *principal components analysis* (PCA). For examples, see Sections 6.10 and 12.2.

6.10 Case Study: Eigenfaces

In computer vision, one discipline is *face recognition.* Suppose that a set of about 1,000 frontal face images is given, and we want to check whether a newly obtained image approximately matches one of the given face images. Each face is represented with a resolution of 100×100 pixels. Column by column, organize the pixels of an image to form a vector; this vector has 10,000 entries. All images thus reside in a 10,000-dimensional linear space F. Each face then corresponds to a vector $\mathbf{v}_i$ in F.

For any arbitrary vector $\mathbf{v}$ (the new face) in F, we would like to know whether it is close to one of the $\mathbf{v}_i$. We could compare $\mathbf{v}$ to each of the 1,000 $\mathbf{v}_i$:

$$\|\mathbf{v}_i - \mathbf{v}\| < \epsilon; \qquad i = 1, \ldots, 1,000 \tag{6.13}$$

for some tolerance ϵ, but that will be expensive. A little preprocessing will speed the process up.

We perform the PCA of the $\mathbf{v}_i$ and keep only the eigenvectors $\mathbf{e}_j$, each with 10,000 components, corresponding to the 100 largest

Figure 6.6. Eigenfaces: three example eigenvectors, re-organized into a three images. (Courtesy of the Center for Ubiquitous Computing (CUbic), Arizona State University.)

eigenvalues.[6] This amounts to saying that each face can be described by 100 features. The vectors $\mathbf{e}_j$—each corresponding to a 100×100 image —are called *eigenfaces.* They do not necessarily look like real faces, but any face in the collection may be approximated by linearly combining them. See Figure 6.6 for an idea of the appearance of eigenfaces.

Now we can approximate each $\mathbf{v}_i$ by a linear combination of the $\mathbf{e}_j$ by solving an overdetermined linear system for each $\mathbf{v}_i$. Then each $\mathbf{v}_i$ is represented by 100 scalars $s_{i,1}, \ldots, s_{i,100}$ (instead of 10,000):

$$\mathbf{v}_i \approx \sum_{j=1}^{100} s_{i,j}\mathbf{e}_j.$$

We denote by $\mathbf{s}_i$ the vector formed by all $s_{i,1}, \ldots, s_{i,100}$. Note that each $\mathbf{s}_i$ has 100 components.

Now we bring in $\mathbf{v}$. We then find the 100 coefficients s_j such that

$$\mathbf{v} \approx \sum_{j=1}^{100} s_j\mathbf{e}_j,$$

again by solving an overdetermined linear system. The vector $\mathbf{s}$ denotes the vector formed by all s_j.

[6]The number 100 is somewhat arbitrary and needs to be refined for a particular data set.

We then check whether $\mathbf{s}$ is close to one of the 1,000 $\mathbf{s}_i$:

$$\|\mathbf{s} - \mathbf{s}_i\| < \epsilon; \qquad i = 1, \ldots, 1,000. \tag{6.14}$$

If it is true for some i, then $\mathbf{v}$ is close to $\mathbf{v}_i$.

To compare the computational cost: (6.13) compares $\mathbf{v}$ against 1,000 vectors $\mathbf{v}_i$ of length 10,000; (6.14) compares $\mathbf{s}$ against 1,000 vectors $\mathbf{s}_i$ of length 100. For real-time face recognition, this is the way to go!

Similar concepts are employed in genomics research, where instead of eigenfaces one computes *eigengenes*. In physics and mechanics, the term *eigenfrequency* is used for systems that will start to vibrate wildly for certain frequencies.

6.11 Singular Values, Volumes, and Determinants

As a practical application, we will use SVD to compute the determinant of a matrix A. We observe that the the matrices V and U in (6.6), being orthonormal, have determinants equal to 1. Thus, the determinant $|A|$ is given by the product of A's singular values:

$$|A| = \sigma_1 \cdot \ldots \cdot \sigma_n. \tag{6.15}$$

Thus, if a 3D object has volume V, it will have volume $\sigma_1\sigma_2\sigma_3 V$ after being transformed by a linear map with singular values σ_1, σ_2, σ_3.

Similarly, a 2D triangle with area F will have area $\sigma_1\sigma_2 F$ after being transformed by a 2D linear map with singular values σ_1, σ_2.

From the definition of the determinant in (6.15), some properties of determinants follow easily:

$$|A^{-1}| = \frac{1}{|A|}, \qquad |AB| = |A||B|, \qquad |A^{\mathrm{T}}| = |A|, \qquad |cA| = c^n|A|.$$

The last of these properties may not be obvious; check for yourself how it follows from (6.15)!

6.12 Problems and Experiments

1. Find a 3×3 matrix with no zero elements and having eigenvalues 3,2,1.

2. In the linear space of all differentiable functions, the derivative operator D (assigning a function f' to a function f) is a linear map. A (nonzero) function u is an *eigenfunction* of D if $Du = \lambda u$; stated differently, $u'(x) = \lambda u(x)$ for all x. Verify that $u(x) = e^{\lambda x} + c$.

3. A Hilbert matrix H is an $n \times n$ matrix with elements

$$h_{i,j} = \frac{1}{i+j-1}; \qquad 1 \le i, j \le n.$$

The 5×5 Hilbert matrix has singular values

$$1.5670, \quad 0.2085, \quad 0.0114, \quad 0.0003, \quad 3.287E-6.$$

It appears that this Hilbert matrix can be approximated reasonably well by a 3×3 matrix. Find this matrix, using (6.12). Then experiment with larger Hilbert matrices.

4. What are examples of 2×2 matrices that have no real eigenvalues?

7

Background: Numerical Calculus

Functions are invaluable tools for scientific computing. Examples include:

- population size as a function of time,
- air pressure on an airplane wing as a function of speed,
- temperature of the atmosphere as a function of CO_2 concentration.

Functions are dealt with in calculus. This chapter presents a brief review.

7.1 Functions

The basic entity in numerical calculus, a *function*, is any procedure that assigns a function value $f(x)$ to a variable, here called x. A good example is the sine function:

$$f(x) = \sin(x).$$

For any x, there can only be *one* function value $f(x)$. We can *plot* this function, thus visualizing its shape. When we plot a function,

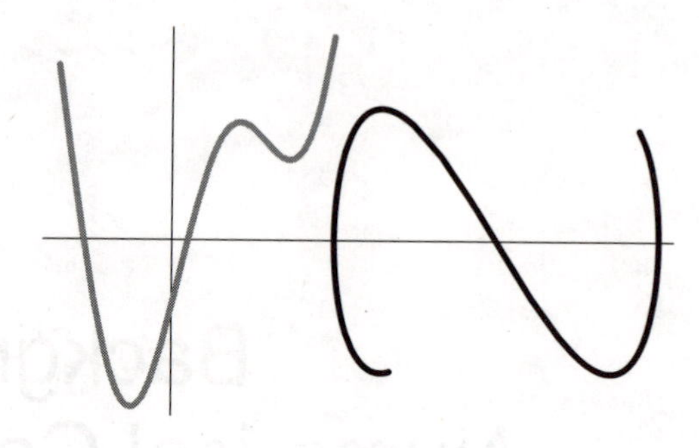

Figure 7.1. Graph of a function (left) and of a nonfunction (right).

technically speaking, we are plotting[1] its *graph*—a set of points (x, y) of the form $(x, f(x))$. Figure 7.1 shows (left) the graph of a viable function and (right) the graph of a nonfunction. The right graph cannot depict a function because both positive and negative y-values are associated with some x-values.

Most functions that we will be dealing with are *continuous*; in effect, this means the graph of the function can be drawn as an uninterrupted line.[2]

Functions are typically not considered over the whole real line, but over an interval $[a, b]$. Even within an interval, a function may not be defined for every point. For example, functions may be defined only over a set of integers. Such functions are called *discrete*.

Functions may have special properties; here we list a handful:

1. *Symmetry.* $f(x) = f(-x)$ for all x. For example, $f(x) = x^2$.

2. *Boundedness.* For all x, $f(x)$ will not exceed a certain bound c, meaning $|f(x)| \leq c$. For example: $f(x) = \cos(x)$.

3. *Monotonicity.* For increasing values of x, the corresponding function values $f(x)$ are also increasing. For example, $f(x) = |x|$ for $x \in [0, 1]$.

4. *Convexity.* No straight line intersects the graph of the function more than twice. For example, $f(x) = x^4$.

[1] More visualization tools are discussed in Chapter 12.

[2] The mathematically precise definition is more involved.

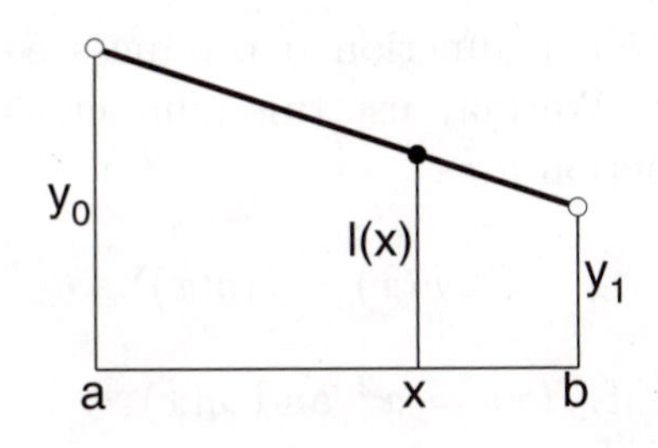

Figure 7.2. Linear interpolation.

We continue our overview of functions with some of the simplest and yet most important ones: the *linear* functions, so called because their graphs are straight lines. They give rise to one of the fundamental computational tools: *linear interpolation.* Suppose we have two function values y_0 and y_1 given at the endpoints of an interval $[a, b]$. Then for any x, its function value $l(x)$ is given by

$$l(x) = \frac{b-x}{b-a} y_0 + \frac{x-a}{b-a} y_1. \tag{7.1}$$

The function $l(x)$ is linear in the variable x, and it also interpolates, meaning

$$l(a) = y_0 \quad \text{and} \quad l(b) = y_1,$$

which can be quickly verified. Figure 7.2 illustrates such a function. Notice that the coefficients in (7.1) sum to one.

Many applications deal with several functions that will need to be combined. We present three ways for combining functions.

1. *Linear combination.* If $f(x)$ and $g(x)$ are two functions defined over the same interval $[a, b]$, then we may define a third function $h(x)$ as

   $$h(x) = \alpha f(x) + \beta g(x),$$

   using two reals α and β that do not depend on x. The fact that we can define h this way leads to the concept of function spaces; see Section 7.5.

2. *Product.* Using the same f and g as above, we may define

   $$h(x) = f(x) \cdot g(x).$$

3. *Composition.* The function g produces a real number $g(x)$ upon input x. We may use this number as input for f, thus creating a function h as

$$h(x) = f(g(x)).$$

For example, if $f(x) = x^2$ and $g(x) = x + 1$, then $h(x) = (x + 1)^2$.

7.2 Limits

A central theme in calculus is that of a *limit.* If we were to evaluate a function, say $f(x) = x^2$, for a particular value of x, we simply plug x into f's definition, to get the answer. But life is not always that easy. Consider, for example,

$$f(x) = \frac{|x|}{x}.$$

This function is rather benign, except at $x = 0$: there it evaluates to $0/0$, which is not a "legal" expression. We therefore study how f behaves near (but not *at*) $x = 0$. That means, we have to look at small positive values for x and, separately, at small negative ones. Let a sequence of positive numbers x approach 0 without quite reaching it, and evaluate f for these arguments. We consistently find $f(x) = 1$. Doing the same for small negative values, we consistently find $f(x) = -1$. More formally, we are taking limits:

$$\lim_{x \to 0+} f(x) = 1 \quad \text{and} \quad \lim_{x \to 0-} f(x) = -1.$$

The first expression—the *right limit*—lets x approach 0 for positive x, the second one—the *left limit*—does the same for negative x. Figure 7.3 shows a plot of f. We can clearly see the jump at $x = 0$. Functions with different left and right limits are not continuous.

A function *has* a limit c at an x-value x_0 if both left and right limits agree. Then we write

$$\lim_{x \to x_0} f(x) = c.$$

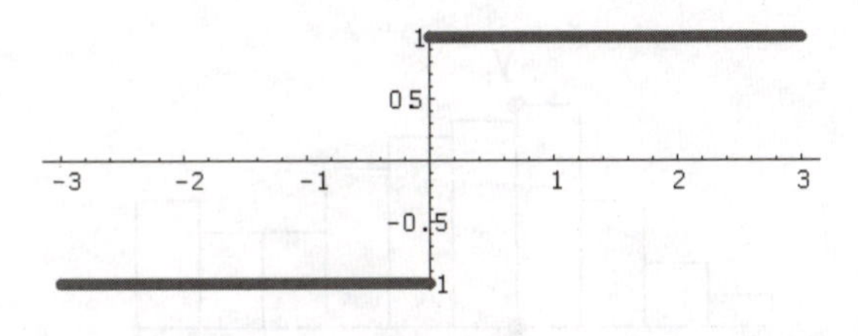

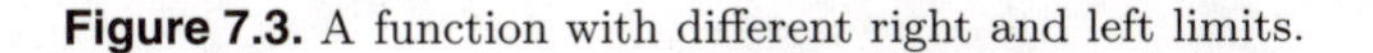

Figure 7.3. A function with different right and left limits.

The most famous limit in all of calculus is the *derivative.* If f is a function, we consider this limit:

$$\lim_{h \to 0} \frac{f(x_0 + h) - f(x_0)}{h}. \tag{7.2}$$

At first sight, this yields the useless expression 0/0, but if f is sufficiently smooth, then the limit exists and is called the derivative $f'(x_0)$. It is useful to remember this as "rise over run," meaning that $f(x_0 + h) - f(x_0)$ measures the change in function value (rise) and h measures the change in the x-direction (run). Figure 7.4 illustrates the derivative as a limit.

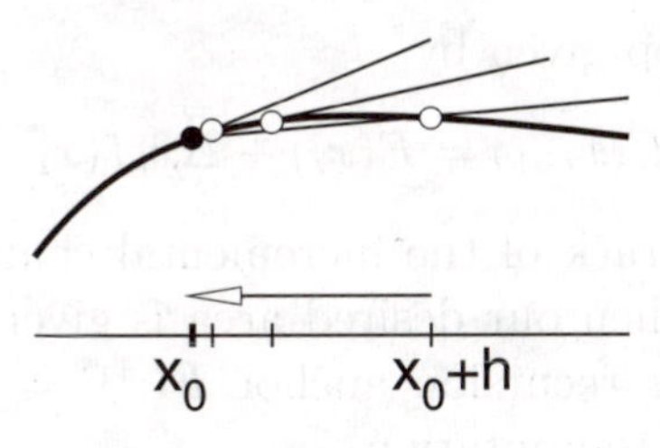

Figure 7.4. The derivative as a limit.

7.3 Integrals

One of the most fundamental tools for a scientist is the *histogram,*[3] which in its simplest form is a block graph of a set of pairs (x_i, y_i). A typical histogram is shown in Figure 7.5.

[3]Histograms are discussed again in Section 12.3

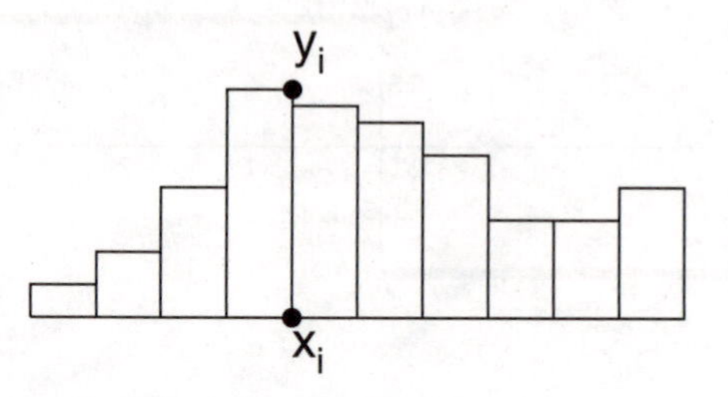

Figure 7.5. A histogram.

It is easy to compute the area A of the histogram:

$$A = \sum_{i=0}^{N-1} \Delta x_i y_{i+1}. \tag{7.3}$$

Here we assume we have x-values $x_0, \ldots, x_N$, and that their spacing is given by $\Delta x_i = x_{i+1} - x_i$. Let us now assume the y_i are computed from a function f, or $y_i = f(x_i)$.

We now introduce a new function F. This function is defined recursively, meaning it has a recursion anchor, in our case given by

$$F(x_0) = c,$$

and a recursive loop given by

$$F(x_{i+1}) = F(x_i) + \Delta x_i f(x_{i+1}). \tag{7.4}$$

That is, F keeps track of the incremental change in area as we increase x-values. Then our desired area is given by $F(x_N) - F(x_0)$, independent of our recursion anchor $F(x_0) = c$. Note that $f(x_0)$ does not enter the computation.

Let's do an example to understand this better. Set $x_0 = 0$ and all $\Delta x_i = 1$. Let $f(x) = x + 1$. With $N = 3$, our function values are $1, 2, 3, 4$. Let's compute the area over x_0, x_3, i.e., $F(x_3) - F(x_0)$. We have

$$\begin{aligned}
F(x_3) &= F(x_2) + f(x_3) \\
&= F(x_1) + f(x_2) + f(x_3) \\
&= F(x_0) + f(x_1) + f(x_2) + f(x_3) \\
&= c + 2 + 3 + 4 \\
&= c + 9.
\end{aligned}$$

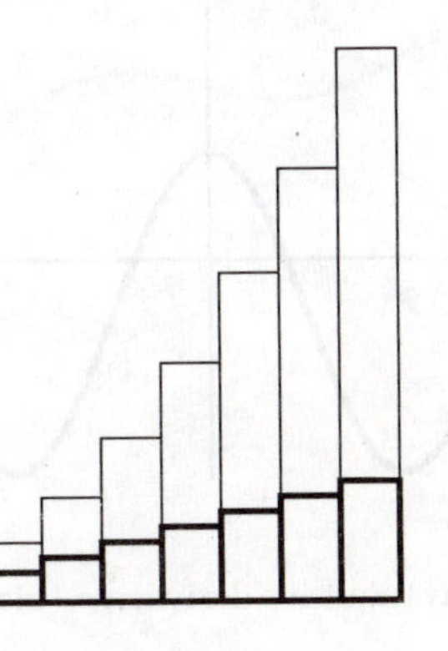

Figure 7.6. Histograms for f (bold) and F (light).

Hence the area is given by $F(x_3) - F(x_0) = 9$. As predicted, it is independent of c! Figure 7.6 shows histograms for the above f and the resulting F, with $N = 7$.

We are next going to increase our sampling rates. More precisely, we will be looking at several histograms, all starting at $x_0 = a$ and ending at $x_N = b$. By increasing the sampling rate we add more x_i into the interval $[a, b]$, or in other words, increase N. As we keep doing this, the histograms for f and for F will more closely approximate these functions. In the limit, as N approaches infinity, we can write (7.3) as the *definite integral*

$$A = \int_a^b f(x)\mathrm{d}x.$$

7.4 Derivatives

Let us take a second look the main (recursive) part (7.4) of the definition of F, and solve for $f(x_{i+1})$:

$$f(x_{i+1}) = \frac{\Delta F(x_i)}{\Delta x_i}$$

where $\Delta F(x_i) = F(x_{i+1}) - F(x_i)$. As the Δx_i become smaller and smaller, this expression approaches $0/0$, and is at first sight an undefined expression. But since we know f exists (we started out with

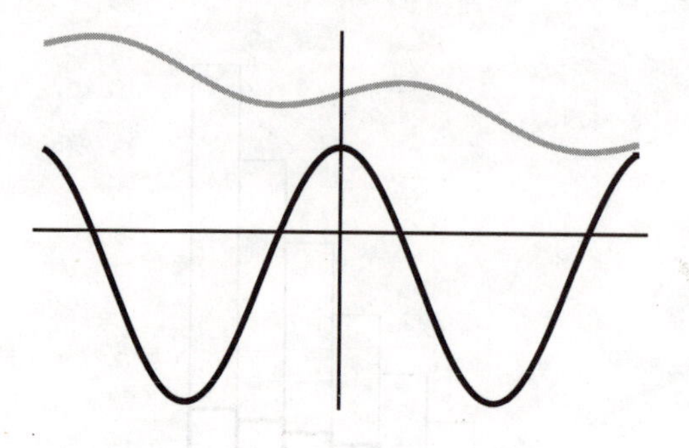

Figure 7.7. A function (gray) and its derivative (black).

it!), the expression does make sense. We write it as

$$f(x) = \frac{\mathrm{d}F(x)}{\mathrm{d}x},$$

and we say that f is the *derivative* of F. Since $\frac{\mathrm{d}F(x)}{\mathrm{d}x}$ exists, we call F *differentiable.* Note that f may not be differentiable itself.

The derivative $g'(x)$ of a function g indicates how much g changes in the vicinity of x. Consider Figure 7.7 for an example. The top function has two local maxima and two local minima. At those points, the function does not change—this is reflected by the derivative being zero at those four locations.

The derivative g' of a function g may be calculated symbolically if an exact equation for the function g is known.[4] We give some simple examples here.

1. *A constant function.* The function $g(x) = c$ does not change at all; it has the same function value c everywhere. Thus its derivative (the rate-of-change indicator) is the zero function: $g'(x) = 0$.

2. *A linear function.* The function $g(x) = ax + b$ has the same rate of change everywhere; it is given by the factor a. The larger a, the larger the slope of the line given by g. We have $g'(x) = a$. This derivative is independent of b, implying that

[4]Sometimes all one has is a procedure for computing g—then numerical methods are called for.

changing b to a different value, thereby translating g along the y–axis, will not affect g'.

3. *The power function.* The function $g(x) = x^n$ has the derivative $g'(x) = nx^{n-1}$.

The derivative of a linear combination of two functions g and h is computed as

$$[\alpha g(x) + \beta h(x)]' = \alpha g'(x) + \beta h'(x). \tag{7.5}$$

This means that the derivative behaves in a linear fashion (in the sense of linear algebra). For example, we compute the derivative of $g(x) = 3 + 3x^5$ as $g'(x) = 15x^4$.

More complicated combinations of functions require more complex rules. For instance, the the *product rule*,

$$[g(x)h(x)]' = g'(x)h(x) + g(x)h'(x),$$

applies for the product of two functions. Let's try this out on $g(x) = x^3$. We can use the product rule if we consider $x^3 = x^2 \cdot x$. Then

$$g'(x) = 2x \cdot x + x^2 \cdot 1 = 3x^2$$

which agrees with the derivative of $g(x)$ using the rule for the power function.

Differentiating a composition of functions $g(h(x))$ calls for the *chain rule:*

$$g(h(x))' = g'(h(x))h'(x).$$

Let's try this out on $g(h(x)) = (x^2)^3$. Here, $h(x) = x^2$. The chain rule gives

$$g(h(x))' = 3(x^2)^2 \cdot 2x = 6x^5,$$

again confirming the power function rule.

So we see that the derivative of a function g is another function g'; we may then take the derivative of the new function g'. This is g's *second derivative*, denoted by g''. This process may be repeated several times (assuming the needed derivative functions do indeed exist). We denote the rth derivative of a function $g(x)$ by $g^{(r)}(x)$;

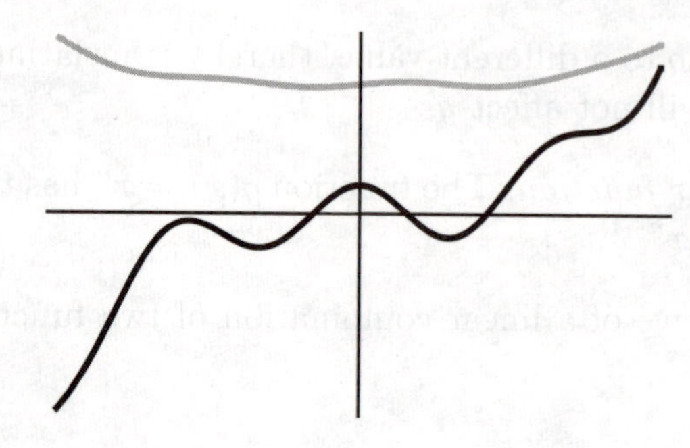

Figure 7.8. A function (gray) and its derivative (black).

for example, $g''(x) = g^{(2)}(x)$. Higher derivatives are important ingredients to methods such as the Taylor expansion; see Section 8.1.

From a practical perspective, we should mention that the process of taking derivatives is a *roughing* process. In Figure 7.8, a function is shown in gray, and its derivative is in black. You clearly see how the derivative accentuates even very slight features in the function. Thus derivatives may be used to detect features in a function.

7.5 Function Spaces

There is a connection between functions and linear spaces. For a simple example, consider the set of all cubic polynomials (called *cubics* for short). Cubics are of the form

$$p(x) = a_0 + a_1x + a_2x^2 + a_3x^3$$

for some real numbers a_0, a_1, a_2, a_3. If we take two cubics

$$p(x) = a_0 + a_1x + a_2x^2 + a_3x^3 \quad \text{and} \quad q(x) = b_0 + b_1x + b_2x^2 + b_3x^3,$$

we may construct another function $\alpha p(x) + \beta q(x)$, where α and β are real numbers that do not depend on x. This function is given by

$$\alpha p(x) + \beta q(x) = (\alpha a_0 + \beta b_0) + (\alpha a_1 + \beta b_1)x + (\alpha a_2 + \beta b_2)x^2 + (\alpha a_3 + \beta b_3)x^3;$$

hence this is another cubic. We just demonstrated that linearly combining cubics yields another cubic; thus, the set of cubics, together

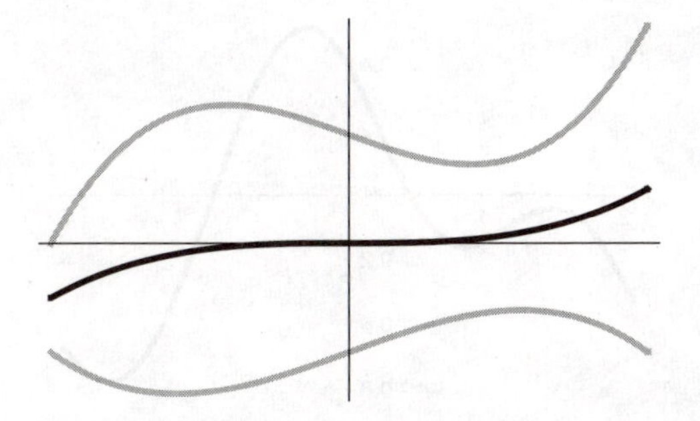

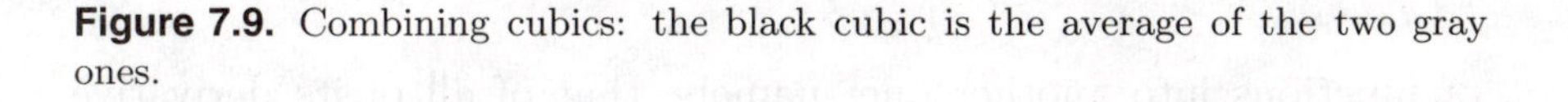

Figure 7.9. Combining cubics: the black cubic is the average of the two gray ones.

with this kind of linear combination, forms a linear space! The elements are not vectors in $\mathbb{R}^n$ anymore, but functions. The zero "vector" is the zero function, obtained by setting all $a_i = 0$. We call this space $\mathbb{P}^3$.

A simple example of combining two cubics with $\alpha = \beta = 0.5$ is shown in Figure 7.9.

What is the *dimension* of $\mathbb{P}^3$? We solve this problem by explicitly naming a basis. One basis is given by the set of functions

$$\{1, x, x^2, x^3\}.$$

Clearly, every cubic may be written as a linear combination of these basis functions, called *monomials*. In addition, the monomials are linearly independent: it is impossible to write one monomial as a linear combination of the others.

This concept generalizes in a fairly obvious way to polynomials of arbitrary degree n, leading to function spaces $\mathbb{P}^n$.

More general function spaces exist. An important one is that of all *continuous functions*, denoted $\mathbb{C}^0$. If f and g are continuous functions, then also $h = \alpha f + \beta g$,[5] thus establishing the linear space condition. Clearly, for any n, $\mathbb{P}^n$ is a subspace of $\mathbb{C}^0$: every polynomial is a continuous function but not vice versa! Other subspaces of $\mathbb{C}^0$ are the spaces of r times differentiable functions, denoted by $\mathbb{C}^r$.

Linear spaces are equipped with linear maps. For function spaces, the most important linear map is the *derivative*: it maps one space

[5]The shorthand notation $h = \alpha f + \beta g$ means $h(x) = \alpha f(x) + \beta g(x)$ for all x.

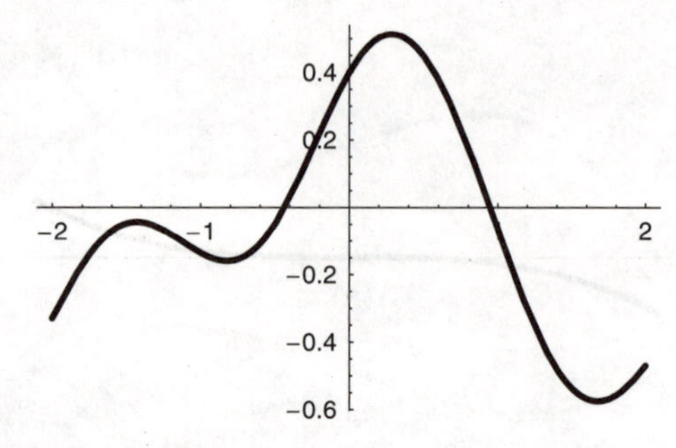

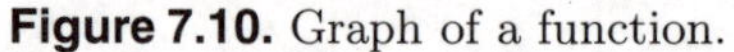
Figure 7.10. Graph of a function.

of functions into another one; namely, that of all of its derivative functions. The map is linear because of (7.5).

7.6 Problems and Experiments

1. Figure 7.10 shows the graph of a function f. Manually sketch its derivative function f'.

2. Again referring to the function f of Figure 7.10, manually sketch its integral function
$$g(x) = \int_{-2}^{x} f(t)\mathrm{d}t; \quad g(-2) = 0.$$

3. Replace f from Figure 7.10 by $f(x) = f(x) + 2$. How will your graphs from problems 1 and 2 change?

4. A process $f(t)$ (for example measuring concentrations of a substance during an experiment) is said to reach *equilibrium* if
$$|f(t) - c| < \epsilon$$
for large values of t and some tolerance ϵ. A simulated process is shown in Figure 7.11. Experiment with plotting several functions until you find one that looks like Figure 7.11.

5. Define a set $\mathcal{F}$ of continuous functions by
$$\mathcal{F} = \{f | f(x) \geq 0\}.$$
Is this set also a linear space of functions? Explain.

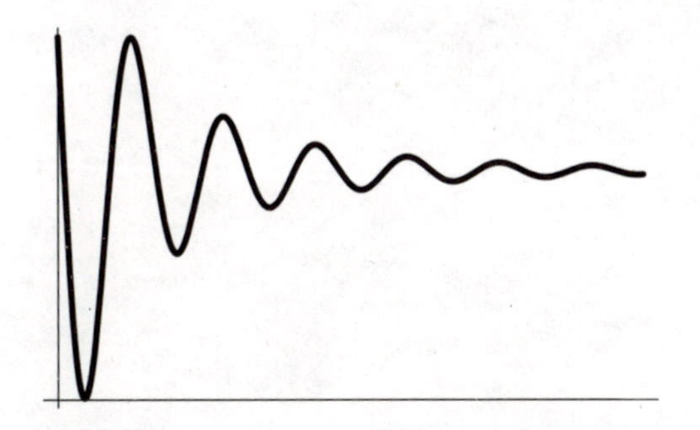

Figure 7.11. Graph of a function.

6. Define three functions f, g, h by $f(x) = \sin^2(x), g(x) = 1, h(x) = \cos^2(x)$. Are these three functions linearly independent? Explain.

8

Data Fitting

Suppose a scientist measures how the volume of an object changes. She records volumes v_i at times t_i. She would then like to plot a graph showing the process—this would be more illustrative than just a table of numbers. Thus, she needs a function that represents the data. There are many ways to find this function; among the available methods are *interpolation* and *approximation*. When we interpolate, we want a function that matches the data exactly; when we approximate, we just want to capture the behavior of the data.

8.1 Taylor Approximation

A famous theorem in calculus states: in the vicinity of an x-value x_0, a function f may be approximated by a polynomial $p(x)$ of degree n by setting

$$p(x) = f(x_0)+(x-x_0)f'(x_0)+\frac{1}{2}(x-x_0)^2f''(x_0)+\ldots+\frac{1}{n!}(x-x_0)^n f^{(n)}(x_0), \tag{8.1}$$

where $f^{(n)}(x_0)$ denotes the nth derivative of f at x_0. This theorem (called Taylor's theorem) is true, but its applicability is restricted to x-values close to x_0. Figure 8.1 illustrates two approximations for the case of $f(x) = \sin(x)$ and $x_0 = 0$, one for $n = 10$ (the "shorter" one) and one for $n = 20$ (the "longer" one). As we increase the degree, we capture more of the function.

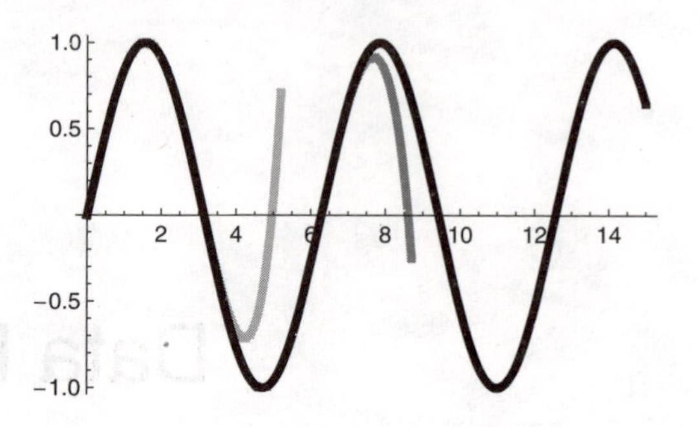

Figure 8.1. Two examples of Taylor approximations (gray) to the sine function (black).

The main value of Taylor's theorem is not for constructing useful approximations, but rather for the design of algorithms that involve derivatives.

8.2 Piecewise Linear Interpolation

Suppose we have a procedure to compute a function $f(t)$ over an interval $a \leq t \leq b$. This may be a function that is very expensive to compute (say, several hours of computer time per function value). Then one way to represent the function is to sample it at some number of arguments $a = t_0 \leq t_1 \leq \cdots \leq t_n = b$ and to construct the polygon $(t_0, f(t_0)), \ldots, (t_n, f(t_n))$; an example of which is shown in Figure 8.2. Depending on the sampling rate, this may be a decent representation of f. In the (theoretical) limit $n \to \infty$, the piecewise linear function will be identical with f.

8.3 Polynomial Interpolation

A different approach is to use polynomials to build an interpolating function. This means we would like to find a polynomial $p(t)$ that interpolates to the data. That is, we would like

$$\begin{aligned} p(t_0) &= v_0, \\ p(t_1) &= v_1, \\ p(t_2) &= v_2, \\ p(t_3) &= v_3. \end{aligned}$$

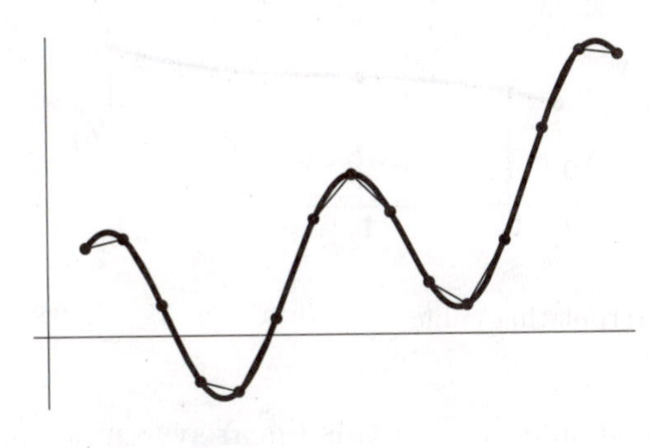

Figure 8.2. Piecewise linear approximation.

Here, we assume measurements are taken at four time steps t_0, t_1, t_2, t_3. These time steps are also referred to as *knots*, a term most commonly used in the context of B-spline methods (introduced in Section 8.6). We now have to be a bit more specific about our polynomial. It should be cubic because then we will have four (unknown) coefficients, which matches the numbers of given data v_i. Thus

$$p(t) = a_0 + a_1 t + a_2 t^2 + a_3 t^3,$$

and we have to find the a_i. For this, we use our given data and write

$$\begin{aligned} v_0 &= a_0 + a_1 t_0 + a_2 t_0^2 + a_3 t_0^3, \\ v_1 &= a_0 + a_1 t_1 + a_2 t_1^2 + a_3 t_1^3, \\ v_2 &= a_0 + a_1 t_2 + a_2 t_2^2 + a_3 t_2^3, \\ v_3 &= a_0 + a_1 t_3 + a_2 t_3^2 + a_3 t_3^3. \end{aligned}$$

These are four equations in as many unknowns. This is begging for matrix notation! We then have

$$\begin{bmatrix} v_0 \\ v_1 \\ v_2 \\ v_3 \end{bmatrix} = \begin{bmatrix} 1 & t_0 & t_0^2 & t_0^3 \\ 1 & t_1 & t_1^2 & t_1^3 \\ 1 & t_2 & t_2^2 & t_2^3 \\ 1 & t_3 & t_3^2 & t_3^3 \end{bmatrix} \begin{bmatrix} a_0 \\ a_1 \\ a_2 \\ a_3 \end{bmatrix}, \tag{8.2}$$

which abbreviates as

$$\mathbf{v} = T\mathbf{a}. \tag{8.3}$$

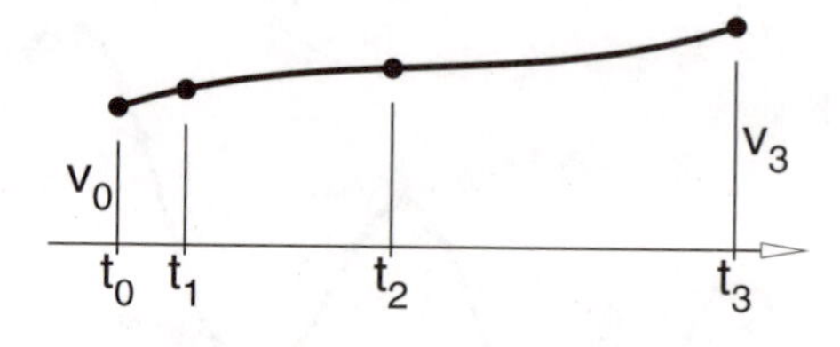

Figure 8.3. An interpolating cubic.

The coefficient matrix T of this linear system is called a *Vandermonde matrix.* But is our linear system solvable, in other words, is T of full rank? If two of the t_i are identical it is not, but that is not likely to happen in any realistic scenario. If all t_i are distinct, then T is indeed of full rank and our linear system has a unique solution. Any of the methods in Chapter 5 will find it. Figure 8.3 shows an example of an interpolating cubic.

Polynomial interpolation is not restricted to the cubic case, however. If we had n pairs $(t_i, v_i); i = 0, \ldots, n$, then we would find an interpolating polynomial

$$p(t) = a_0 + a_1 t + \ldots + a_n t^n$$

simply by solving a system with $n+1$ equations for the $n+1$ unknowns a_i, in complete analogy to the cubic case.

A word of caution: polynomial interpolation is simple, but it is dangerous. Figure 8.4 shows the result of interpolating to $(0,0)$, $(1,1)$, $(4,2)$, $(9,3)$, $(16,4)$ with a fourth-degree polynomial. The given data come from the function $\sqrt{t}$, plotted in gray. The interpolating polynomial is shown in black. The result is correct, but for all practical purposes it is disastrous! So if you use polynomial interpolation in some package, be aware that the results may not be what you intended.

8.4 Polynomial Least Squares Approximation

We have seen that using polynomials for interpolation can be tricky, but using polynomials for *approximation* is very reliable. The scenario is now slightly different from that in Section 8.3. We are given a relatively large number of data $(t_i, v_i); i = 0, \ldots, L$, and we would

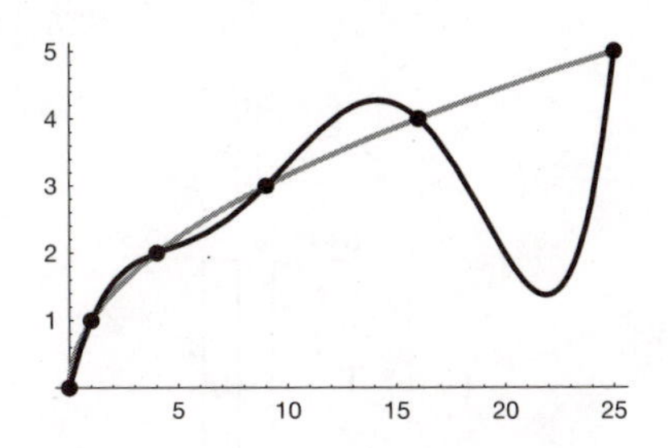

Figure 8.4. An interpolating degree 4 polynomial (black).

like to approximate it by a polynomial $p(t)$ of degree less than L, typically much less.

Without elaborating on what we mean by "to approximate," we would most likely want the data points to be close to the approximating polynomial p. Ideally, p would pass through the data points:

$$p(t_0) = v_0,$$
$$\vdots$$
$$p(t_L) = v_L.$$

Assuming the polynomial to be

$$p(t) = a_0 + a_1 t + \ldots + a_n t^n,$$

this becomes

$$p(t_0) = a_0 + a_1 t_0 + \ldots + a_n t_0^n,$$
$$\vdots$$
$$p(t_L) = a_0 + a_1 t_L + \ldots + a_n t_L^n,$$

which can be written matrix form as

$$\mathbf{v} = T\mathbf{a}. \tag{8.4}$$

At first sight, this looks exactly like (8.3), but there is a difference: here, we have an *overdetermined* linear system with more equations than unknowns. In (8.3), we had the same number of knowns and unknowns.

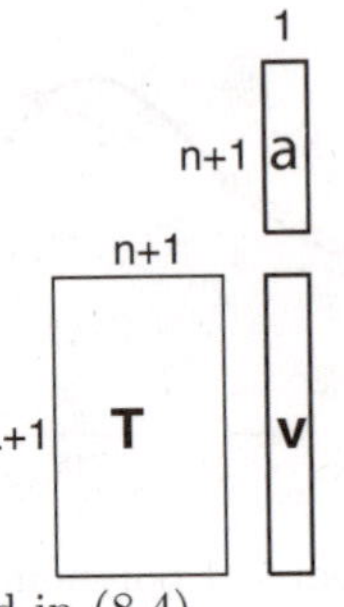

Figure 8.5. The matrices involved in (8.4).

The matrix T (sometimes referred to as the *design matrix*) has more rows than columns; a sketch of the dimensions is given in Figure 8.5. We multiply both sides of (8.4) by T^{T} and obtain

$$T^{\mathrm{T}}\mathbf{v} = T^{\mathrm{T}}T\mathbf{a}. \tag{8.5}$$

This linear system, known as the *normal equations*, has a square and symmetric coefficient matrix $T^{\mathrm{T}}T$. The coefficient matrix is invertible, and we may solve (8.5) by using Gauss elimination (see Section 5.3) or any other method. (We encountered overdetermined systems in Section 5.8 as well.) However, if L is large, $T^{\mathrm{T}}T$ can be ill-conditioned (i.e., close to being singular), and then singular value decomposition (see Section 6.6) is more reliable.

The polynomial p is known as the *least squares polynomial* approximating the given data. It is optimal in the following sense. One measure for the error E between p and the data values is given by

$$E = \sum_{i=0}^{L}[v_i - p(t_i)]^2.$$

This is the *least squares error.*[1] Among all polynomials, p is the one for which E is the smallest. In many cases, the data values v_i will be afflicted by "noise"; the least squares approach is designed to deal with this efficiently. But sometimes it is known that some data values have larger noise levels than others. In that case, it is feasible to weight the equations; we replace the equation

$$p(t_i) = a_0 + a_1 t_i + \ldots + a_n t_i^n$$

[1]This error essentially measures the average deviation of p from the data.

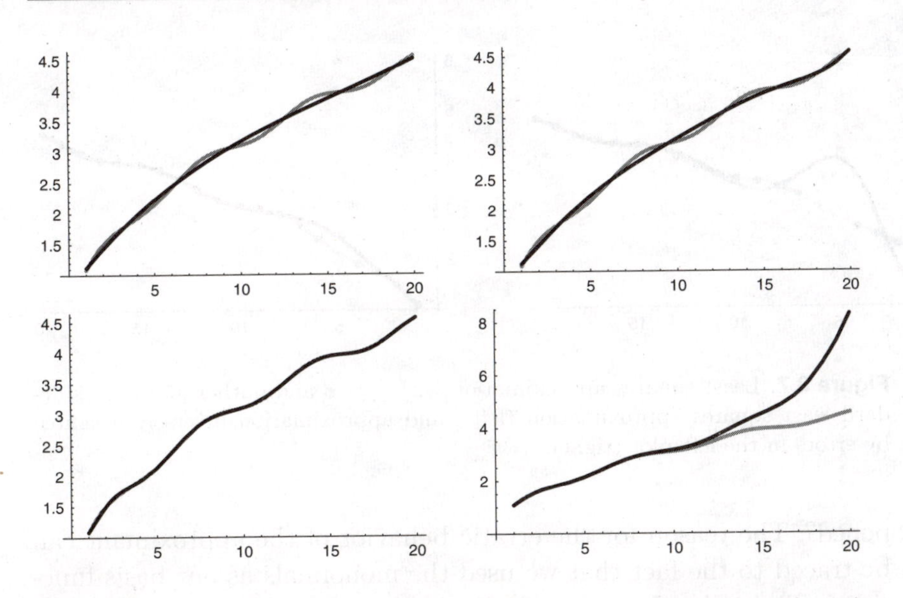

Figure 8.6. Least squares approximations. For various values of n: $n = 3$ (top left); $n = 6$ (top right); $n = 10$ (bottom left); $n = 11$ (bottom right). The function is shown in gray, and the approximation is black.

by

$$w_i p(t_i) = w_i a_0 + w_i a_1 t_i + \ldots + w_i a_n t_i^n,$$

where w_i is given a large value if our confidence in the value v_i is high, and low otherwise. Although multiplying the left- and the right-hand sides of an equation by the same value does not seem to achieve much, it will affect the least squares solution. In particular, giving a huge value to only few of the w_i will force $p(t_i)$ to come arbitrarily close to the corresponding v_i.

We finish this discussion with some examples. Let's pick the function

$$f(t) = \sqrt{t} + 0.1 \sin t$$

to approximate. The sine function here uses radians for its argument. Let's also pick 20 values $t_0 = 1, \ldots, t_{19} = 20$. Figure 8.6 shows approximants of degrees 3, 6, and 10; the original function f is in gray, the approximants are in black. As expected, the quality of the approximations improves with higher degrees.

But now something ugly happens. If we increase the degree to $n = 11$, we get the bottom right result of Figure 8.6! What hap-

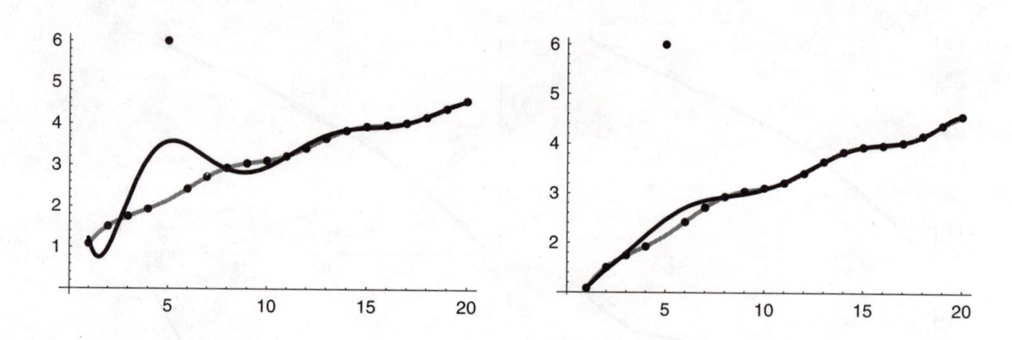

Figure 8.7. Least squares approximation with $n = 8$ and outlier at $t = 5$: standard least squares approximation (left), and approximation inversely weighted by errors in the left plot (right).

pened? The reason for the erratic behavior of the approximant can be traced to the fact that we used the monomials as our basis functions. These simply become difficult to evaluate for values as large as $t = 20$. In more technical terms, we see that the monomial basis is *unstable* away from $t = 0$. This is reflected by the condition numbers of the matrices $T^{\mathrm{T}}T$. (See Section 6.7 for a definition of condition number.) For $n = 11$, the condition number is 10^{33}, making the problem intractable. For a remedy, see Section 8.5.

The weighted least squares approach is useful for dealing with *outliers*, any data values that are in crass disagreement with the general trend. We first compute the standard least squares fit p, and then compute weights

$$w_i = \frac{1}{|p(t_i) - v_i| + 0.1}. \tag{8.6}$$

These are then used in a weighted least squares computation. The constant 0.1 is heuristic; it simply ensures that no division by 0 will happen. See Figure 8.7 for an example. If desired, this method may be repeated.

8.5 General Least Squares Approximation

Basis functions other than monomials may be used. Then our least squares problem becomes: given a set $\{t_i, v_i\}, i = 0, \ldots, L$, find a function

$$f(t) = c_0 B_0(t) + \ldots + c_n B_n(t)$$

that best approximates the given data in the least squares sense. The functions B_i could be polynomials, trigonometric functions, or exponential functions, as long as they are linearly independent.

Proceeding exactly as in Section 8.4, we obtain an overdetermined linear system of equations:

$$\mathbf{v} = T\mathbf{c}. \tag{8.7}$$

Here, the matrix T is obtained by evaluating the basis functions B_j at the t_i:

$$T = \begin{bmatrix} B_0(t_0) & \cdots & B_n(t_0) \\ \vdots & & \vdots \\ \vdots & & \vdots \\ B_0(t_L) & \cdots & B_n(t_L) \end{bmatrix}.$$

The solution is again via the normal equations, or, in tricky cases, via singular value decomposition.

We now show how to fix the problem with the monomial degree 11 fit of Figure 8.6. As our new basis functions, we pick the *Bernstein polynomials* B_j^n:

$$B_j^n(t) = \binom{n}{j} \frac{(t_L - t)^{n-j}(t - t_0)^j}{(t_L - t_0)^n}; \quad j = 0, \ldots, n. \tag{8.8}$$

Using this basis, the matrix $T^{\mathrm{T}}T$ has, for $n = 11$, the condition number 10^6. Compare this to 10^{33} for the monomial basis! Needless to say, the degree 11 fit is no longer a problem anymore.

8.6 B-Spline Interpolation

In addition to polynomials, another class of functions is widely used for data fitting problems: these are *spline functions* in B-spline form. B-splines are functions that consist of polynomial segments instead of just one polynomial. Assume our data are given over an interval $[a, b]$. A *partition* of this interval is an increasing set of values

$a = u_0, \leq u_1, \ldots, \leq u_K = b$. The partition is also referred to as a *knot sequence* and the u_i are referred to as *knots.* A spline function consists of cubic polynomial segments that are different for each interval $[u_i, u_{i+1}]$ in such a way that they agree in all derivatives up to order 2.[2] For a fixed partition, the set of all splines forms a linear space.

A spline $s(t)$ may be written as a linear combination of *basis functions* $N_i^3(t)$:

$$s(t) = \sum_{i=0}^{K+2} N_i^3(t) d_i. \tag{8.9}$$

The basis functions are defined recursively:

$$N_i^n(t) = \frac{t - u_{i-1}}{u_{i+n-1} - u_{i-1}} N_i^{n-1}(t) + \frac{u_{i+n} - t}{u_{i+n} - u_i} N_{i+1}^{n-1}(t). \tag{8.10}$$

This recursion is anchored by the definition

$$N_i^0(t) = \begin{cases} 1 & \text{if} \quad u_{i-1} \leq t < u_i, \\ 0 & \text{otherwise.} \end{cases} \tag{8.11}$$

In order for this to work, we need to add several "phantom knots" such that our knot sequence becomes $u_0, u_0, u_0, u_0, u_1, \ldots, u_{K-1}, u_K, u_K, u_K, u_K$. This is strictly a technicality and shall not concern us here. We should note that the N_i^3 do depend on the knot sequence $\{u_i\}$.

Figure 8.8 shows six B-splines over a *uniform* knot sequence, meaning the u_i are equally spaced. Figure 8.9 shows the effect of a different knot sequence.

B-splines have many valuable properties; we list several here:

1. *Partition of unity.* For every t, all B-splines add to 1:

$$\sum_{i=0}^{K+2} N_i^3(t) \equiv 1.$$

[2]More general definitions are possible, but we will keep things simple here.

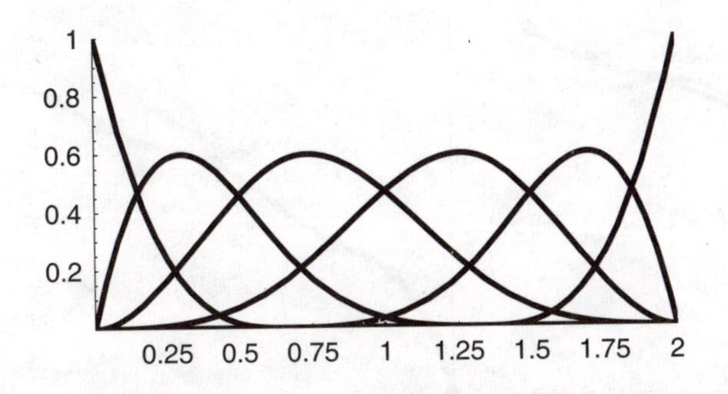

Figure 8.8. The cubic B-splines $N_0^3, \dots N_5^3$ over the knot sequence $0, 0.66, 1.33, 2$.

2. *Local support.* The B-spline $N_i^3(t)$ is nonzero only over the interval $[u_{i-3}, u_{i+1}]$. Over that interval, it is also nonnegative.

3. *Polynomial inclusion.* For the special case $K = 1$, the B-splines are polynomials—in particular, they are the cubic Bernstein polynomials (8.8).

4. *Linear precision.* The linear function $y = t$ may be written as a spline curve:

$$t = \sum_{i=0}^{K+2} \frac{u_i + u_{i+1} + u_{i+2}}{3} N_i^3(t).$$

5. *Linear independence.* Every spline has a *unique* representation in terms of B-splines.

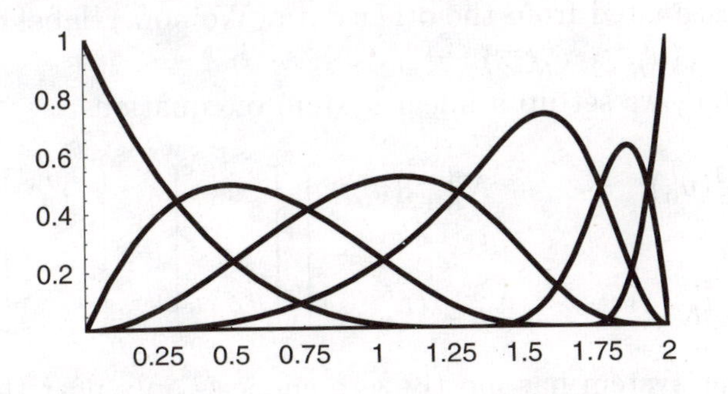

Figure 8.9. The cubic B-splines $N_0^3, \dots N_5^3$ over the knot sequence $0, 1.33, 1.7, 2$.

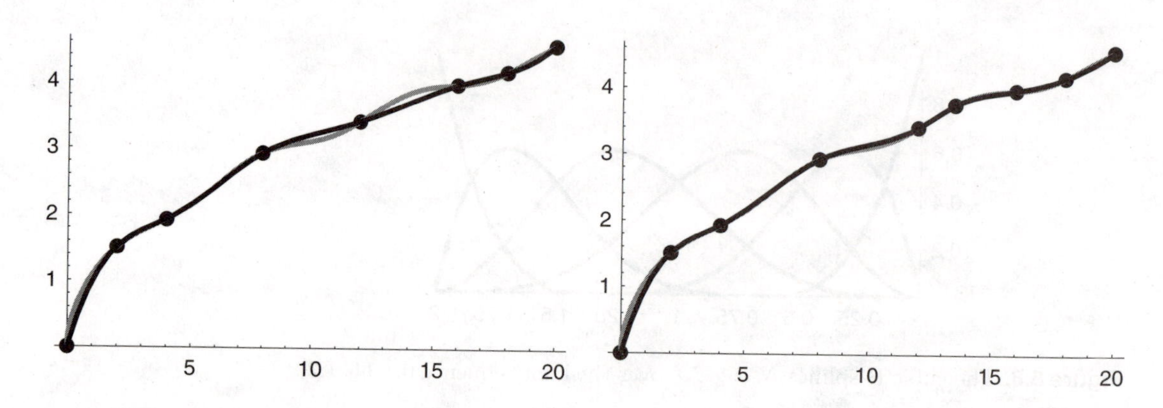

Figure 8.10. Spline interpolation over knot sequence 0, 2, 4, 8, 12, 16, 18, 20 (left), and after adding one more data value at $t = 13.5$ (right).

The typical spline interpolation problem is posed as follows:

Given: A set of data points (u_i, v_i); $i = 0, \dots, K$

Wanted: Coefficients d_i such that the resulting spline $s(t)$ interpolates:

$$s(u_i) = v_i; \quad i = 0, \dots, K. \tag{8.12}$$

Our spline is of the form (8.9), and we see that we have a problem: there are $0, \dots, K+2$ unknowns but only $0, \dots, K$ interpolation conditions (8.12); we are short two conditions! The standard way to combat this problem is to supply two extra conditions, such as requiring interpolation at two more locations $(u_0 + u_1)/2$ and $(u_{K-1} + u_K)/2$. The corresponding function values can (with some luck) be estimated from the other data. We now relabel our data to $(u'_0, v'_0), \dots, (u'_{K+2}, v'_{K+2})$.

As usual, we set up a linear system of equations:

$$\begin{bmatrix} N_0^3(u'_0) & \cdots & N_{K+2}^3(u'_0) \\ \vdots & & \vdots \\ N_0^3(u'_{K+2}) & \cdots & N_{K+2}^3(u'_{K+2}) \end{bmatrix} \begin{bmatrix} d_0 \\ \vdots \\ d_{K+2} \end{bmatrix} = \begin{bmatrix} v'_0 \\ \vdots \\ v'_{K+2} \end{bmatrix} \tag{8.13}$$

This linear system has mostly zero entries—only near the diagonal are there nonzero entries. Such systems typically behave in a very

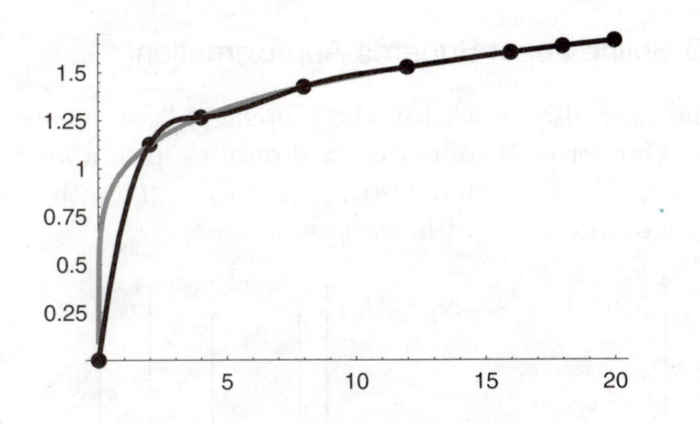

Figure 8.11. Spline interpolation: unwanted oscillations.

stable way. If we compare the condition number of this interpolation approach to that of polynomial interpolation, the difference is striking. For example, the coefficient matrix corresponding to the data from Figure 8.10 has condition number 5.0!

The left graph of Figure 8.10 shows an example of a cubic spline interpolant with knot sequence 0, 2, 4, 8, 12, 16, 18, 20. As you can see, the interpolant is too far from the data around $t = 13$. Adding another function value at $t = 13.5$ resolves the problem; see the right graph of Figure 8.10.

But even spline interpolation has its problems: Figure 8.11, the data suggest a convex curve, but the spline exhibits some oscillations, which are most likely unwanted. Even adding more data is no safe cure for situations like this. More specialized methods exist to handle such cases, but they are beyond the scope of this book.

Finally, we must include a word on the "trick" of adding data at $u'_1 = (u_0 + u_1)/2$ and $u'_{K+1} = (u_{K-1} + u_K)/2$. What are the corresponding function values v'_1 and v'_{K+1}? One way to estimate them is by fitting a quadratic to the three data pairs $(u_0, v_0), (u_1, v_1), (u_2, v_2)$ and to evaluate that quadratic at u'_1. This gives the desired v'_1. The value v'_{K+1} is found analogously.

Many algorithms do not estimate additional function values at the two ends, but instead estimate slopes at u_0 and u_K. Those slopes are then entered into the linear system—we do not pursue this kind of *end condition* here.

8.7 B-Spline Least Squares Approximation

B-splines may also be used in the context of least squares approximation. Our set of B-splines is fixed once we pick a knot sequence $u_0, \ldots, u_K$. We are given data $(t_i, v_i); i = 0, \ldots, L$ (with $L > K$) and obtain an overdetermined linear system:

$$\begin{bmatrix} N_0^3(t_0) & \cdots & N_{K+2}^3(t_0) \\ \vdots & & \vdots \\ \vdots & & \vdots \\ N_0^3(t_L) & \cdots & N_{K+2}^3(t_L) \end{bmatrix} \begin{bmatrix} d_0 \\ \vdots \\ d_{K+2} \end{bmatrix} = \begin{bmatrix} v_0 \\ \vdots \\ \vdots \\ v_L \end{bmatrix}. \tag{8.14}$$

We solve this for the unknown $d_0, \ldots, d_{K+2}$ exactly as we did in the polynomial least squares setting. B-splines are a particularly nice choice for basis functions for the following reason: the coefficient matrix in (8.14) is *sparse*, meaning that most entries are 0. To see why, simply take any row i of the matrix. Its entries are the values

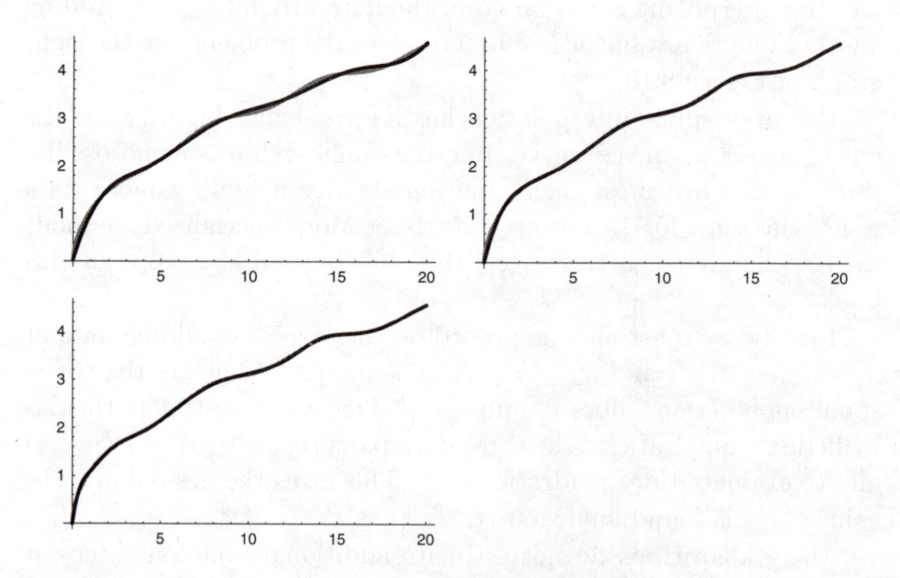

Figure 8.12. B-spline least squares approximations with $u_i = 0, 4, 8, 12, 16, 20$ (top left); $u_i = 0, 2, 4, 6, 8, 12, 14, 16, 18, 20$ (top right); and the addition of knot $u_1 = 1$ (bottom).

for all B-splines at t_i. But at t_i, most of these B-splines will be zero! In fact, at most four (cubic) B-splines will be nonzero at any t_i.

Now let's look at some examples. Again, we use the function

$$f(t) = \sqrt{t} + 0.1 \sin t,$$

where the sine function uses radians for its argument. We pick 20 uniform values $t_0 = 1, \ldots, t_{19} = 20$. We first try the uniform knot sequence 0, 4, 8, 12, 16, 20. The resulting least squares fit is shown in Figure 8.12 (top left). We improve our result by increasing the elements in the knot sequence to 0, 2, 4, 6, 8, 10, 12, 14, 16, 18, 20, which produces Figure 8.12 (top right). This fit is much improved, but is still not perfect near $t = 0$. (You'll have to look hard to spot this!) We add one more knot at $t = 1$, and now we obtain an even better result, shown in Figure 8.12 (bottom).

Typically, the more knots in the knot sequence, the better the fit. But how many knots? As a rule, if we have about 2 to 5 knots in each interval $[u_i, u_{i+1}]$, then we should get a decent fit without using too many knots.

8.8 Integrals and Derivatives

Integrals and derivatives were introduced in Sections 7.3 and 7.4, respectively. In most practical cases, integrals and derivatives are not computed symbolically because many times this is impossible or too tedious; rather, a numerical approach is taken. For both cases, a standard approach is to use some data fitting technique and then to perform the desired operation based on it.

Let's start with integration, the more benign process of the two that we cover in this section. We saw how to approximate a function f by a piecewise linear function in Section 8.2. The approximating polygon consists of points (x_i, y_i); $i = 0, \ldots, N$, where $y_i = f(x_i)$. We may form a *trapezoid* T_i for any two subsequent x_i, x_{i+1}, given by the four points

$$(x_i, 0), \quad (x_i, y_i), \quad (x_{i+i}, y_{i+1}), \quad (x_{i+1}, 0).$$

The area A_i of T_i is given by

$$A_i = (x_{i+1} - x_i) \frac{y_i + y_{i+1}}{2}.$$

The sum of all these areas is an approximation of the area formed by f over the interval $[x_0, x_N]$ (see Figure 8.13). This is a fairly robust method, and much more efficient than the histogram approach of Section 7.3.

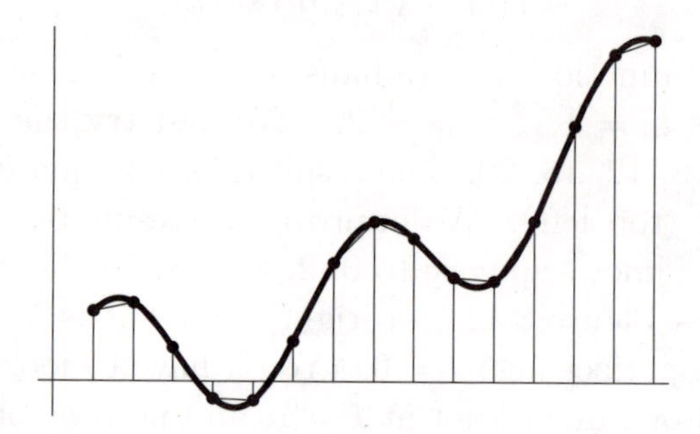

Figure 8.13. The trapezoid rule for integration.

Simpson's rule uses piecewise quadratic interpolation instead of piecewise linear interpolation. Consider three consecutive values x_{i-1}, x_i, x_{i+1} with equal spacing h and their corresponding function values. These three data points may be fitted by a quadratic polynomial as shown in Section 8.3 and Figure 8.14. Explicitly integrating that polynomial gives

$$S_i = \frac{h}{3}[y_{i-1} + 4y_i + y_{i+1}]$$

for the area under the polynomial from x_{i-1} to x_{i+1}. If N is even, then there are $N/2$ areas S_i. Their sum is the desired integral.

Simpson's rule is more costly to compute than the trapezoidal rule, but the benefit is higher accuracy.

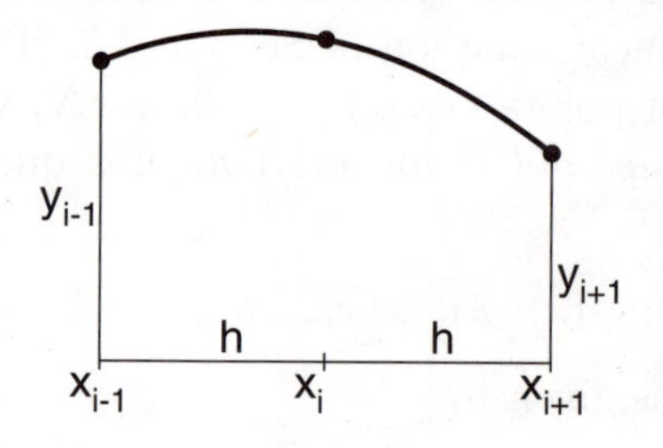

Figure 8.14. Simpson's rule.

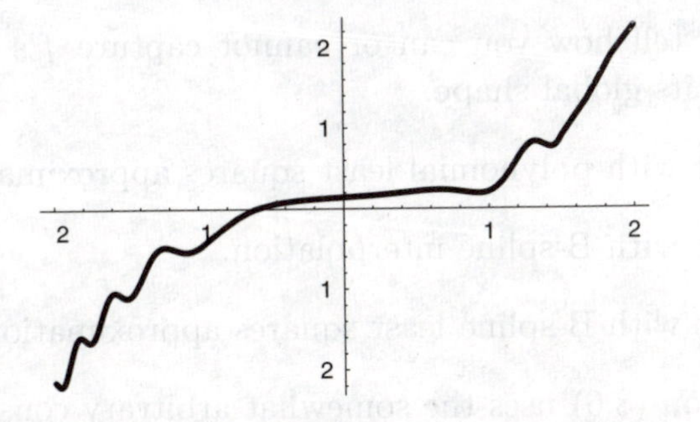

Figure 8.15. Graph of function $f(x) = \frac{1}{10}x + \frac{1}{7}\cos(3x^3) + \frac{1}{4}x^3; \ -2 \le x \le 2.$

Let us now turn to derivatives, which pose more numerical difficulties. Suppose we have function values y_{i-1}, y_i, y_{i+1} at x_{i-1}, x_i, x_{i+1}, respectively. How can we use these to estimate a derivative at x_i? Let us assume the x-values are equally spaced with distance h, and furthermore that h is small. A decent approximation for the slope at x_i is then given by

$$y_i' = \frac{y_{i+1} - y_{i-1}}{2h}.$$

In theory, the smaller the h, the better the estimate. But for very small h, we are close to $y_i' = 0/0$, and that needs to be avoided!

Sometimes second derivatives y_i'' are needed. We reuse Figure 8.14. We compute the quadratic polynomial through y_{i-1}, y_i, y_{i+1}. This time, we differentiate it twice and find

$$y_i'' = \frac{y_{i+1} - 2y_i + y_{i-1}}{h^2}.$$

8.9 Problems and Experiments

For several of the following problems, we will use the function

$$f(x) = \frac{1}{10}x + \frac{1}{7}\cos(3x^3) + \frac{1}{4}x^3; \qquad -2 \le x \le 2,$$

which is graphed in Figure 8.15.

1. Experiment with interpolating polynomials to recapture f. Use various degrees and knots. Comment on your efforts. In par-

ticular tell how you can or cannot capture f's shape details versus its global shape.

2. Repeat with polynomial least squares approximation.

3. Repeat with B-spline interpolation.

4. Repeat with B-spline least squares approximation.

5. Equation (8.6) uses the somewhat arbitrary constant 0.1. Experiment with different choices.

6. Experiment with combatting outliers using B-spline least squares.

9

Computing Dynamic Processes

A dynamic process is a phenomenon that changes its attributes over time. Most dynamic processes are modeled by using *ordinary differential equations* (ODEs), which were invented to model the rate of change of a phenomenon. There are two ways to attack ODEs: symbolically or numerically. For most real-life problems, there are no symbolic solutions, and thus we concentrate on numerical methods.

9.1 Background

We start with a simple example. One of the oldest models to study *population growth* goes back to T. Malthus around 1800. Malthus was interested in predicting human population size over many years. He invented a simple model, which states that population grows at a rate proportional to the existing population.[1] If $p(t)$ denotes population at a given time t, then $p'(t)$ is the rate of change at that time—for example, if $p'(t) = 0$, then population does not change at time t. Thus

$$p'(t) = c \cdot p(t), \tag{9.1}$$

where some constant c depends on the population at hand (since some grow faster than others). If we know the population at some

[1] This is not terribly realistic, and much more sophisticated and complicated models exist now.

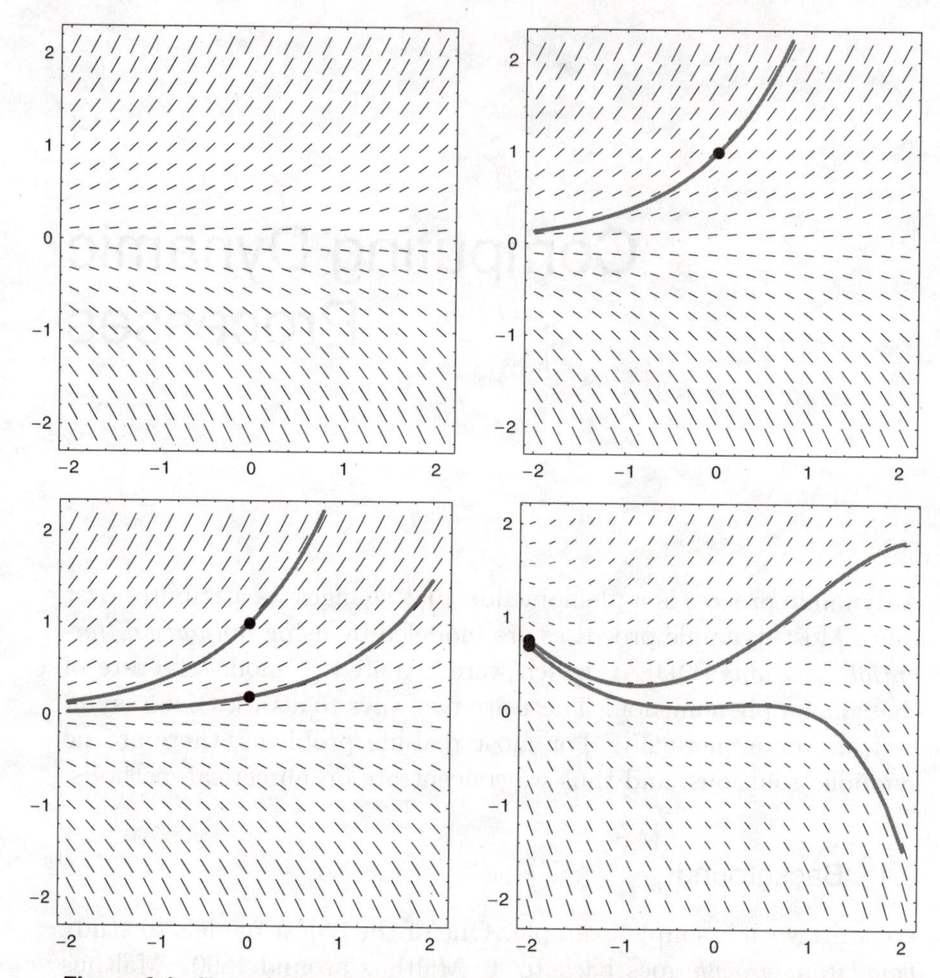

Figure 9.1. Slope fields and ODEs, for which the initial value is a filled circle. A slope field (top left) is shown with a solution defined by an initial value (top right) and two different solutions defined by two different initial values (bottom left). For a different slope field, two very close initial values may result in very different solutions (bottom right).

time t_0 (called the *initial value*), then (9.1) allows us to compute the population at any other time t—we'll soon see how. Because we start from a known value p_0, this type of ODE is also known as an *initial value problem.*

Now consider a coordinate system with horizontal axis t and vertical axis p. For every pair (t, p), (9.1) gives a *slope* $p'(t)$. This slope (recall: slope=rise/run) may be plotted as a short line segment having slope $p'(t)$. Higher slopes result in steeper line segments. To visualize what is going on here, let us plot our short line segments at a finite array of points (t_i, p_j). We get an image similar to the one in Figure 9.1 (top left). There, we picked $c = 1$ in (9.1). Note that for constant p, all slopes are identical![2]

How is this slope field related to solving (9.1)? Figure 9.1 (top right) illustrates that if we pick any starting point, such as $(0, 1)$ in the figure, there is a unique function through that point having slopes determined by the slope field. If we pick another starting point, then that same slope field will give us another function. Figure 9.1 (bottom left) shows two different starting points and the corresponding functions.

Intuitively, then, if we have a slope field generated by an ODE and if we have a starting point, there is a function through it with the required slopes. Next, we turn this observation into algorithms.

In (9.1), only the first-order derivative p' is used; thus, this is called a *first-order ODE.* If derivatives of order r are involved, we speak of rth-order ODEs.

Of course, most ODEs will not be as simple as (9.1). Let's look at another example: $p'(t) = p - 0.4t^2$. Figure 9.1 (bottom right) shows the corresponding slope field and two solutions corresponding to different initial values. Note here how close the initial values are and how different the corresponding solutions are!

9.2 Euler's Method

Let us now label our coordinate system by using the standard x, y notation. An ODE associates a slope $y'(x)$ with every point (x, y), just as we saw in Figure 9.1. This slope (in general) is described by a function f of x and y:

$$y'(x) = f(x, y). \tag{9.2}$$

In our example (9.1), we had

$$y'(x) = c \cdot y(x);$$

[2]This is true for this example, but not in general.

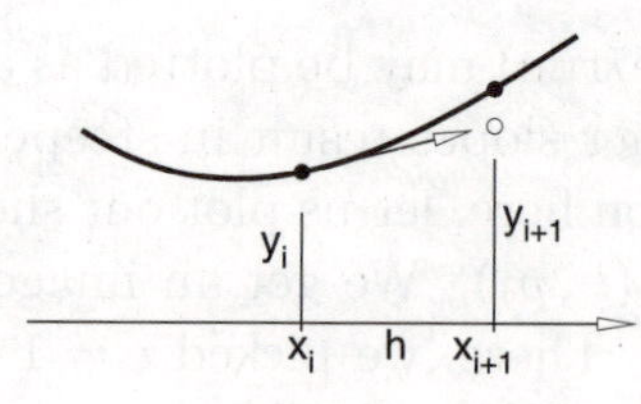

Figure 9.2. Euler's method: starting from (x_i, y_i), we move in the tangent direction to find (x_{i+1}, y_{i+1}). Note how the true value (solid circle) is missed by the approximation (hollow circle).

thus, $f(x, y) = c \cdot y(x)$. We also have one function value: $y(x_0)$. The idea is to trace the solution in the slope field by stepping along the x-axis and computing appropriate approximations to $y(x)$. We step using a *step length h*; this generates a sequence $x_0, x_1, x_2, \ldots$, with

$$x_i = x_0 + ih. \tag{9.3}$$

We then have to compute the corresponding y-values, $y_i = y(x_i)$.

At the starting point (x_0, y_0), we know the function value y_0 and also the derivative $y'(x_0) = f(x_0, y_0)$ from (9.2). We approximate $y(x)$ by its tangent at x_0 and use this approximation to compute $y(x_1)$. The tangent at x_0 is a linear function of x and is given by

$$t(x) = y(x_0) + f(x_0, y_0) \cdot (x - x_0).$$

Choosing a "reasonably small" value for h, we step to $x_1 = x_0 + h$ from the relationship in (9.3). Next, evaluate at x_1 and use the result as an approximation for $y(x_1)$:

$$y(x_1) = y(x_0) + hf(x_0, y_0).$$

Continuing on, we have

$$y_{i+1} = y_i + hf_i, \tag{9.4}$$

where we set $f_i = f(x_i, y_i)$. We have generated a sequence of points (x_i, y_i), which ideally are samples from our unknown function $y(x)$. Figure 9.2 shows the geometry behind the construction.

Now it's time for an example! Let's take the ODE $y' = y$ with initial value $y(0) = 1$.[3] We are interested in finding values for y over

[3]Of course, we have an exact solution $y = e^x$ for this simple ODE, but we will not make use of that fact.

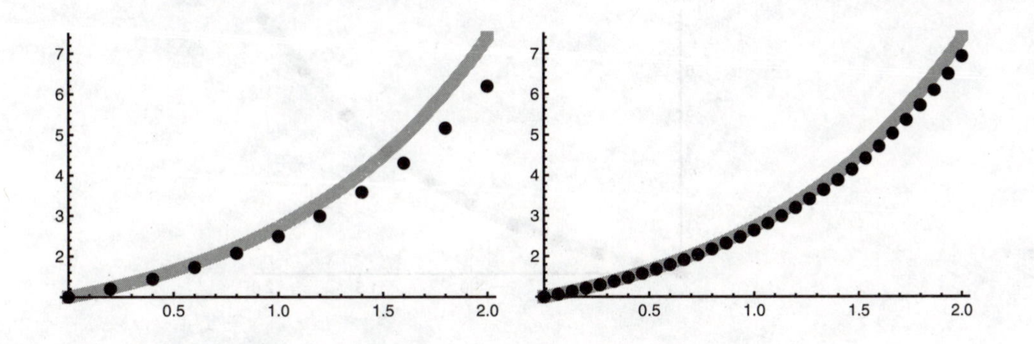

Figure 9.3. Euler's method, using 10 steps (left), and using 30 steps (right). The exact solution is shown in gray.

the interval $[0, 2]$. Figure 9.3 shows what happens if we run Euler's method with 10 and 30 steps, respectively. The ten-step result is very poor; increasing the number of steps from 10 to 30 (thus using a smaller h) significantly improves the solution, although it is clearly not perfect.

It turns out that Euler's method is simply too crude to be of practical value. However, it does nicely illustrate the principle of approximating a solution to an initial-value problem.

9.3 Heun's Method

Euler's method is a *predictor* method: at each x_i, it predicts a new value y_{i+1}, uses it, and then proceeds from there. We have seen that the prediction is not always wonderful, and some more caution might be called for. That is what *predictor-corrector* methods do: they inspect the newly found y_{i+1} and use that information to rerun the prediction step.

A simple example is *Heun's method*. It computes y_{i+1} from the Euler step (9.4). It then computes the slope f_{i+1} at (x_{i+1}, y_{i+1})—but not to carry out the next Euler step! Instead, it readjusts the slope f_i by averaging it with f_{i+1}. Thus a Heun step is given by

$$y_{i+1} = y_i + \frac{1}{2}h(f_i + f_{i+1}). \tag{9.5}$$

We test for our simple example $y' = y$, using 10 points in the interval $[0, 2]$ and an initial value $y(0) = 1$. The result is shown in

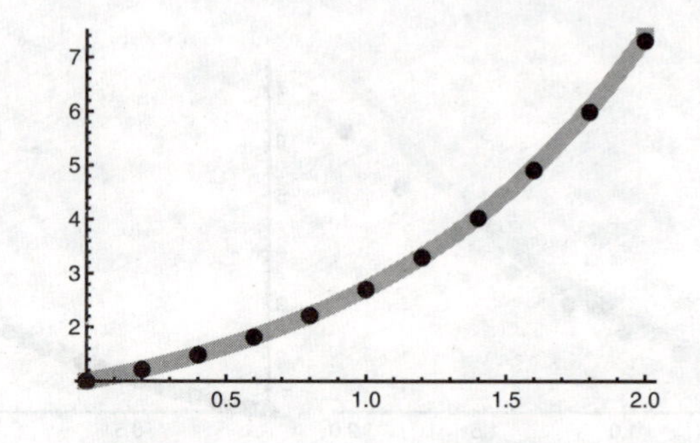

Figure 9.4. Heun's method with 10 steps. Exact solution: gray.

Figure 9.4. It is nearly perfect compared to that shown in Figure 9.3 using Euler's method.

For another comparison, we use the function $f(x, y) = y - 0.4x^2$. Figure 9.1 (bottom right) shows two solutions corresponding to initial values $(-2, 0.7)$ and $(-2, 0.76)$. Let's use the second one and see how Heun's and Euler's solutions compare with 50 sampled points. As Figure 9.5 shows, Heun's method (top) generates a set of points very close to the solution, but Euler's method (bottom) gets lost in the slope field!

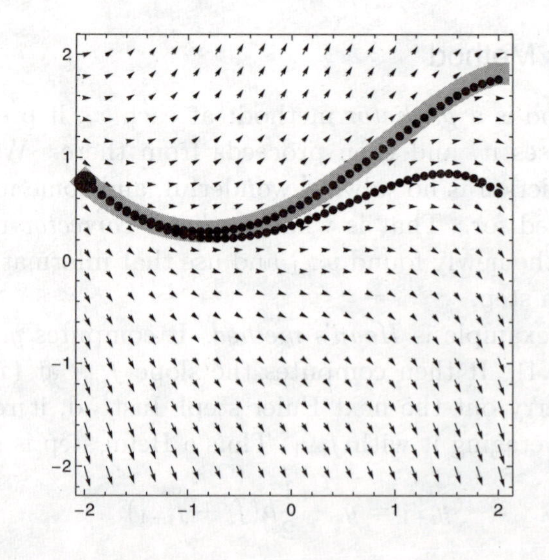

Figure 9.5. Heun's method (top) versus Euler's method (bottom). The exact solution is shown in gray.

9.4 Boundary Values

We have so far been concerned with ODEs of the form

$$y'(x) = f(x, y).$$

More general ODEs exist that involve higher derivatives. The next level up thus looks like this:

$$y''(x) = g(x, y, y'),$$

with an example being

$$y''(x) = y + y' - x^2.$$

We concentrate here on *linear* ODE's having the form

$$y''(x) = p(x)y'(x) + q(x)y(x) + r(x). \tag{9.6}$$

Instead of prescribing an initial value $y'(x_0)$, we now prescribe two values:

$$y(a) = y_0 \qquad \text{and} \qquad y(b) = y_n$$

for some integer n. The result of the *finite difference method* is a sequence of points $(x_0, y_0), (x_1, y_1), \ldots, (x_n, y_n)$. As usual, we space our x-values uniformly: $x_{i+1} - x_i = h$.

We utilize the following two approximations for first and second derivatives:

$$y'(x_i) \approx \frac{y_{i+1} - y_{i-1}}{2h}, \qquad y''(x_i) \approx \frac{y_{i-1} - 2y_i + y_{i+1}}{h^2}. \tag{9.7}$$

Inserting these approximations into (9.6) gives

$$\frac{y_{i-1} - 2y_i + y_{i+1}}{h^2} = p_i \frac{y_{i+1} - y_{i-1}}{2h} + q_i y_i + r_i. \tag{9.8}$$

As usual, subscripts i denote evaluation at x_i.

We multiply through by h^2:

$$y_{i-1} - 2y_i + y_{i+1} = hp_i \frac{y_{i+1} - y_{i-1}}{2} + h^2 q_i y_i + h^2 r_i$$

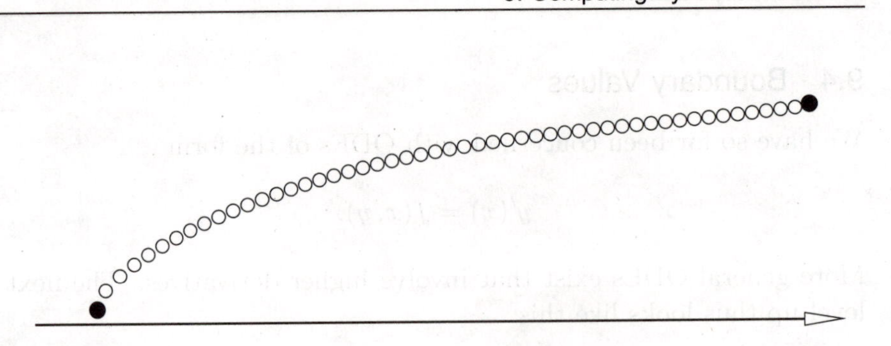
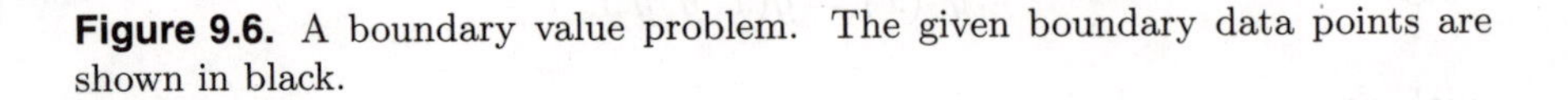

Figure 9.6. A boundary value problem. The given boundary data points are shown in black.

and rearrange to get

$$(1+\frac{h}{2}p_i)y_{i-1}-(2+h^2q_i)y_i+(1-\frac{h}{2}p_i)y_{i+1} = h^2r_i; \qquad i = 1,\ldots,n-1. \tag{9.9}$$

For the first ($i = 1$) and last ($i = n - 1$) equations, y_0 and y_n are known and will be moved to the right-hand side. We then have $n - 1$ unknowns $y_1, \ldots, y_{n-1}$, and there are $n - 1$ equations in (9.9); thus, we have a linear system for the y_i. Using the abbreviations

$$a_i = 1 + \frac{h}{2}p_i, \qquad b_i = -(2 + h^2q_i), \qquad c_i = 1 - \frac{h}{2}p_i,$$

we have the matrix form

$$\begin{bmatrix} b_1 & c_1 & & & & \\ a_2 & b_2 & c_2 & & & \\ & a_3 & b_3 & c_3 & & \\ & & & \ddots & & \\ & & & a_{n-2} & b_{n-2} & c_{n-2} \\ & & & & a_{n-1} & b_{n-1} \end{bmatrix} \begin{bmatrix} y_1 \\ y_2 \\ y_3 \\ \vdots \\ y_{n-2} \\ y_{n-1} \end{bmatrix} = \begin{bmatrix} h^2r_1 - a_1y_0 \\ h^2r_2 \\ h^2r_3 \\ \vdots \\ h^2r_{n-2} \\ h^2r_{n-1} - c_{n-1}y_n \end{bmatrix}.$$

We arrived at a *linear* system of equations because we started out with a linear ODE. For nonlinear ODEs, we would have to solve a nonlinear system of equations. Unfortunately, even in the linear case, a solution does not always exist; obviously, this depends on the behavior of the functions p, q, r.

The basic geometry of a boundary value problem is illustrated in Figure 9.6.

9.5 ODEs and Dynamical Systems

In Section 9.1, we looked at slope fields. Now, we will generalize that concept to *vector fields.* A vector field assigns a vector $[v(x,y), w(x,y)]^{\mathrm{T}}$ to every point $[x,y]^{\mathrm{T}}$ in the x,y-plane. As an example, we may use a vector field to describe the velocity vector of a hurricane at any given point in its path. Figure 9.7 shows an image of Hurricane Katrina. Velocity vectors follow the spiraling-inward behavior of hurricanes.

Figure 9.8 shows a vector field. One way of interpreting it is as an image of a fluid in motion.

In Section 9.1, we studied how a simplified population model led to an ODE. We will now look at a much more sophisticated model, one that models not one population, but two; this is the so-called predator-prey model. To make things somewhat intuitive,

Figure 9.7. Hurricane Katrina. (Figure courtesy of NOAA.)

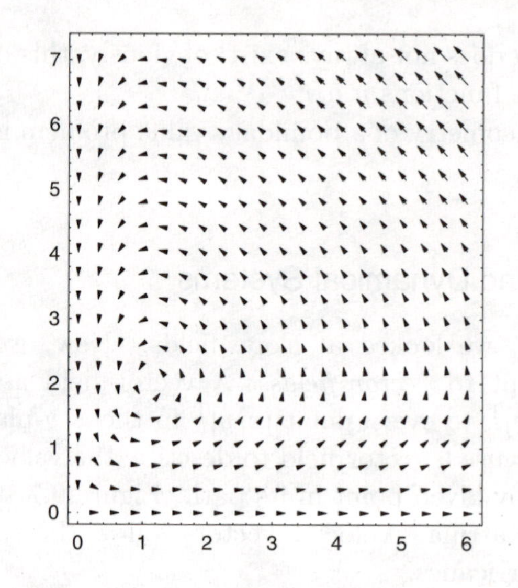

Figure 9.8. A vector field.

we'll consider one population of foxes and one of rabbits. The sizes of these populations are interrelated in an intriguing way. For example, many foxes will decimate the rabbits; on the other hand, a small number of rabbits will result in fewer foxes due to lack of food.

The following system of ODEs (describing a *dynamical system*) makes this more precise. At time t, we will have $f(t)$ foxes and $r(t)$ rabbits. Here is a first model for how they interact:

$$r'(t) = \quad r(t) - f(t)r(t), \tag{9.10}$$

$$f'(t) = \quad -f(t) + f(t)r(t). \tag{9.11}$$

Rabbits eat grass, and so as long as they do not fall victim to the foxes, they multiply. Hence the $r(t)$ term in (9.10). The $-f(t)r(t)$ models how the foxes diminish the rabbit population. Any time a fox and a rabbit cross paths, the rabbit gets eaten. This may happen in $f(t) \cdot r(t)$ ways.

As for (9.11), the $-f(t)$ term reflects the fact the fox numbers will decrease without rabbits around. However, if there are rabbits, the $f(t)r(t)$ term, while decreasing rabbit numbers, helps the fox numbers.

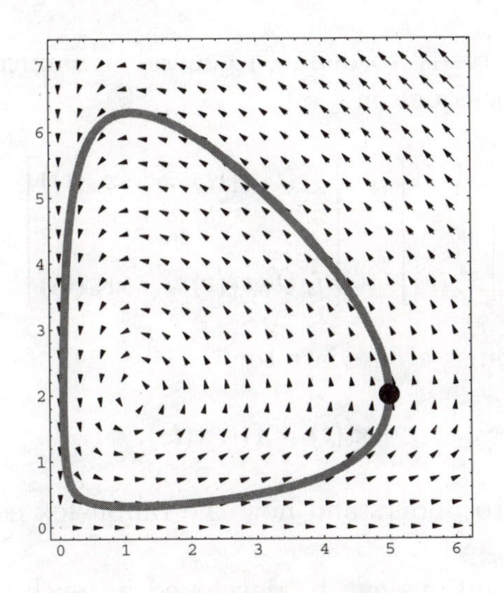

Figure 9.9. A solution to the predator-prey problem.

A more precise model, known as the Lotka-Volterra model, uses additional constants; we selected suitable constants, resulting in

$$r'(t) = \quad 2r(t) - 0.5f(t)r(t), \tag{9.12}$$
$$f'(t) = \quad -f(t) + 0.9f(t)r(t). \tag{9.13}$$

This is the vector field of Figure 9.8. A solution needs a starting situation: for some time $t = 0$, the numbers $f(0)$ of foxes and $r(0)$ of rabbits. Figure 9.9 shows a solution starting with $r(0) = 5$, $f(0) = 2$. The horizontal axis shows the number of rabbits; the vertical one shows the number of foxes.

Having a solution is important, but how do we interpret it? At the beginning, we have more rabbits than foxes; that means plenty of food for the foxes—their numbers grow. This is reflected in the solution curve that grows in the f-direction as you follow the arrows in the vector field. At some point, the number of foxes reaches a peak, but now they have decimated the rabbits to an extent that starvation triggers a steep decline in fox numbers. Consequently, the number of rabbits increases, and we return to our initial state and start all over again. If the solution exhibits this kind of stability, it is known as a *stable orbit*.

We are now ready to define a general dynamical system. It is given by a vector equation

$$\begin{bmatrix} x_1'(t) \\ \vdots \\ x_n'(t) \end{bmatrix} = \begin{bmatrix} f_1(t, x_1(t), \ldots, x_n(t)) \\ \vdots \\ f_n(t, x_1(t), \ldots, x_n(t)) \end{bmatrix}, \tag{9.14}$$

which we sometimes abbreviate to

$$\mathbf{x}'(t) = \mathbf{f}(t, \mathbf{x}(t)). \tag{9.15}$$

It is important to understand how the rabbit-fox equations fit this mold!

Numerical solutions can be developed in much same way as for ODEs. To illustrate the principle, we give the example of Euler's method. We have initial values $\mathbf{x}(t_0)$ at t_0 and plan to find a sequence of points $\mathbf{x}(t_i)$ (or for short, $\mathbf{x}_i$) that approximates the solution for arguments t_i. The derivative $\mathbf{x}'(t)$ of a vector function $\mathbf{x}(t)$ at t_i may be approximated by

$$\mathbf{x}_i' = \frac{\mathbf{x}_{i+1} - \mathbf{x}_i}{h}$$

for some small value h. Recalling $\mathbf{x}_i = \mathbf{f}(t, \mathbf{x}_i)$ (or for short, $\mathbf{f}_i$), we have

$$\mathbf{f}_i = \frac{\mathbf{x}_{i+1} - \mathbf{x}_i}{h}$$

and thus

$$\mathbf{x}_{i+1} = \mathbf{x}_i + h\mathbf{f}_i.$$

Note the similarity of this equation with (9.4)! Although Euler's method is not the most accurate, it should show that "normal" ODEs and systems of ODEs have similar numerical solution strategies.

9.6 Case Study: The Lorenz Attractor

In 1963, E. Lorenz set up a system of ODEs to model the behavior of air over heated terrain. If $x(t), y(t), z(t)$ describe the path of a particle, then these functions satisfy a nonlinear, coupled system of

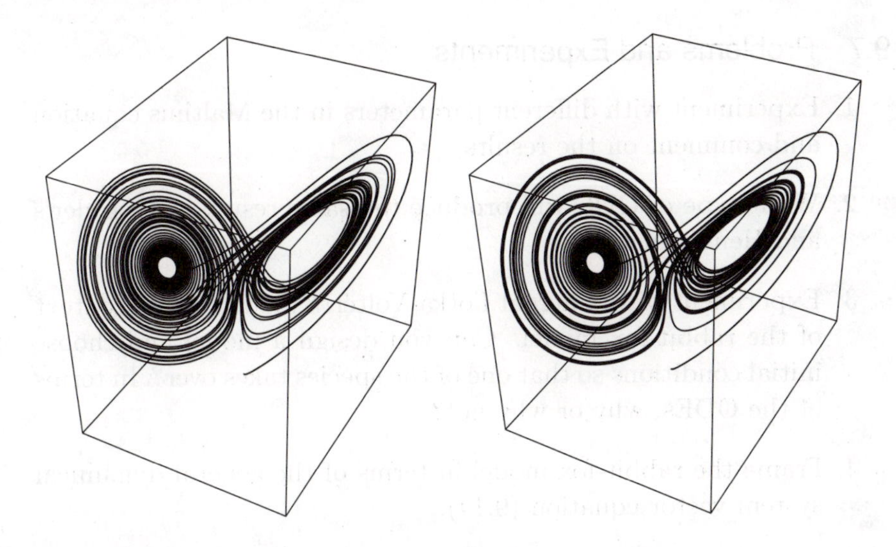

Figure 9.10. The Lorenz attractor for $y_0 = 1.0$ (left) and for $y_0 = 1.00001$ (right).

ODEs:

$$\begin{aligned} x'(t) &= a(y(t) - x(t)), \\ y'(t) &= x(t)(b - z(t)) - y(t), \\ z'(t) &= x(t)y(t) - cz(t). \end{aligned}$$

The constants a, b, c are physical constants. A realistic choice is $a = 10, b = 28, c = 8/3$. Figure 9.10 shows what happens for two different initial conditions. Both examples in the figure start with $x_0 = z_0 = 0$. Their initial y-values, however, are ever so slightly different.

At first sight, the two resulting trajectories in Figure 9.10 look the same, but upon closer inspection, we see they are not identical at all. A tiny change in initial conditions caused very different trajectories—a behavior that is typically referred to as *chaotic*.[4] Neither trajectory is stable: the particle moves from one orbit to the other but never settles in. The Lorenz attractor can look stunningly beautiful—check out the Internet for images, including applets!

[4]This is related to the "butterfly effect": a butterfly flapping its wings could (in theory) cause a tornado in another part of the world.

9.7 Problems and Experiments

1. Experiment with different parameters in the Malthus equation and comment on the results.

2. What type of ODE will produce the same results with Euler's and Heun's methods?

3. Experiment with different Lotka-Volterra models in the context of the rabbit-fox model. Can you design a model and choose initial conditions so that one of the species takes over? In terms of the ODEs, why or why not?

4. Frame the rabbit-fox model in terms of the general dynamical system vector equation (9.14).

5. Experiment with the coefficients in the Lorenz attractor ODEs.

10

Finding Roots

A simulation of a chemical process might involve a substance f increasing during a process, and a substance g decreasing during the same process. The process should be terminated once both substances have reached the same level. Both substance concentrations are time dependent, so $f = f(t)$ and $g = g(t)$. Equilibrium is reached at the time t when $f(t) = g(t)$. Setting

$$y(t) = f(t) - g(t),$$

we are now looking for the root of $y(t)$. The term "root" refers to the zero-crossing of the function y, that is, to finding the value t for which

$$y(t) = 0.$$

Of course, there could be more than one of those! For most "real-life" situations, y's zeroes will have to be found approximately, that is to say, to the precision required.

10.1 The Piecewise Linear Approach

Here is a "quick-and-dirty" way for finding the roots of a function $y(x)$. Simply replace (approximate) y by a polygon $\{x_i, y_i\}$, i.e., by a piecewise linear approximation, as shown in Figure 10.1.

Typically, an interval $[a, b]$ for a likely zero is known, and an easy choice for the x_i is to distribute them evenly within the interval $[a, b]$.

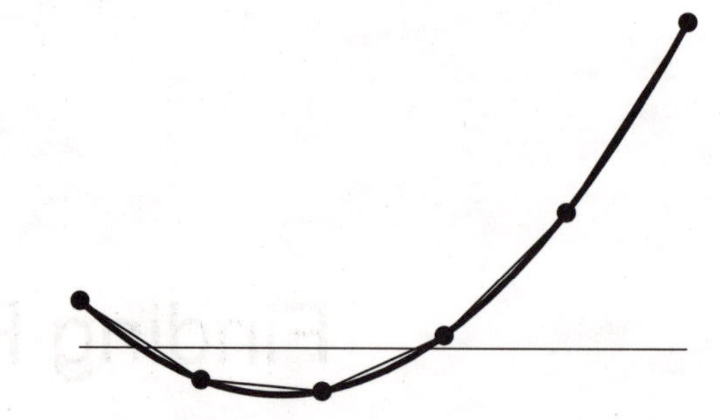

Figure 10.1. A piecewise linear approximation of a function.

Then we check each polygon leg to see whether it crosses the x-axis: this happens if $y_i \cdot y_{i+1} < 0$, meaning there is a sign change in the function values. Then the zero $\hat{x}_i$ is easily found from

$$\frac{-y_i}{\hat{x}_i - x_i} = \frac{y_{i+1} - y_i}{x_{i+1} - x_i}.$$

This simple identity expresses the slope (rise/run) of the line segment (x_i, y_i), (x_{i+1}, y_{i+1}) in two different ways; see Figure 10.2.

While simple, this method is also dangerous: it may require a large number of function evaluations.[1] It may also miss a zero, as shown in Figure 10.3.

In many practical situations data are obtained from measurements (t_i, y_i). If the piecewise linear data of Figure 10.3 are given and zeroes need to be found, it is up to the skills of a user to make an educated decision.

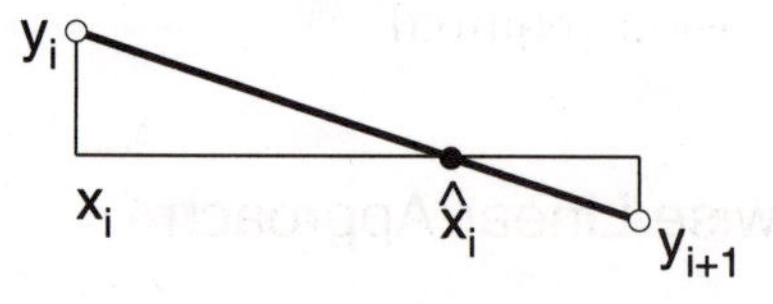

Figure 10.2. Finding the zero of a line segment.

[1]Keep in mind that "function evaluation" might mean running a simulation of a complex process, which can be extremely costly!

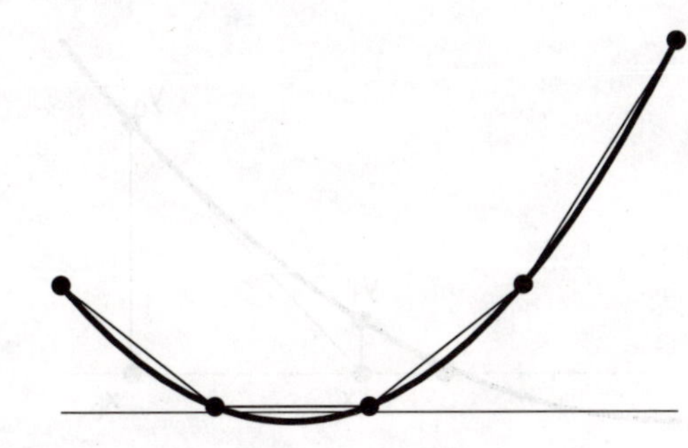

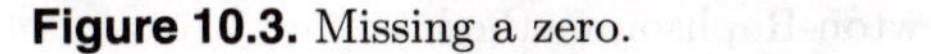

Figure 10.3. Missing a zero.

10.2 The Newton-Raphson Method

Let's suppose we have an educated guess about where a function $y(x)$ crosses the x-axis. If the guess is indeed good, then the Newton-Raphson technique will produce a highly accurate solution iteratively.

Let the guess x-value be x_0. Typically, $y(x_0) \neq 0$, and an improvement is called for. The solution involves locally replacing y by its tangent. The slope y'_0 of this tangent is given by

$$y'_0 = \frac{y_0}{x_0 - x_1}, \tag{10.1}$$

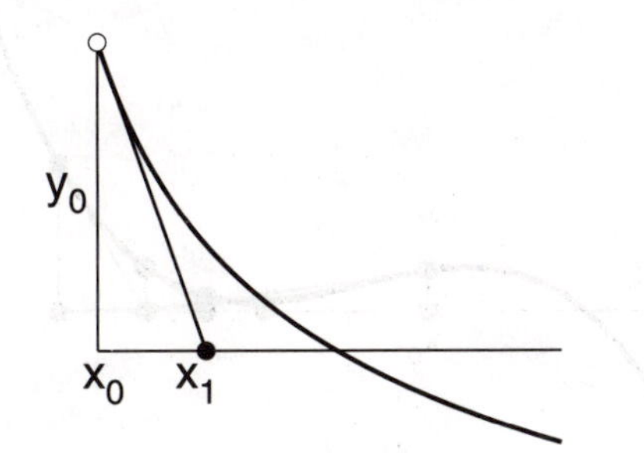

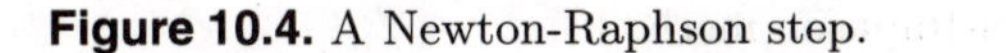

Figure 10.4. A Newton-Raphson step.

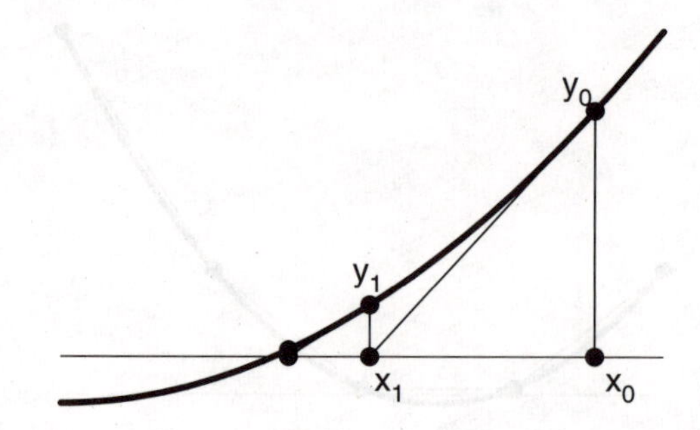

Figure 10.5. Some iterations of the Newton-Raphson method.

where x_1 denotes the tangent's zero crossing. The value x_1 should be closer to y's true zero than x_0. We find it from (10.1):

$$x_1 = x_0 - \frac{y_0}{y_0'}. \tag{10.2}$$

Figure 10.4 illustrates this step.

Having found x_1, we continue this process, producing x-values of $x_2, x_3, \ldots$ until we are close enough to the desired zero. Figure 10.5 illustrates some iterations.

We do have to discuss a big caveat in our approach, though: x_0 is supposed to be "close" to the desired zero. This is not guaranteed, so we may find ourselves in situations such as that shown in Figure 10.6. You might want to trace the individual steps!

As a rule, x_0 should be close enough to the desired zero z such that y is monotone between z and x_0. If this is the case, convergence is extremely fast, although, again it is not guaranteed!

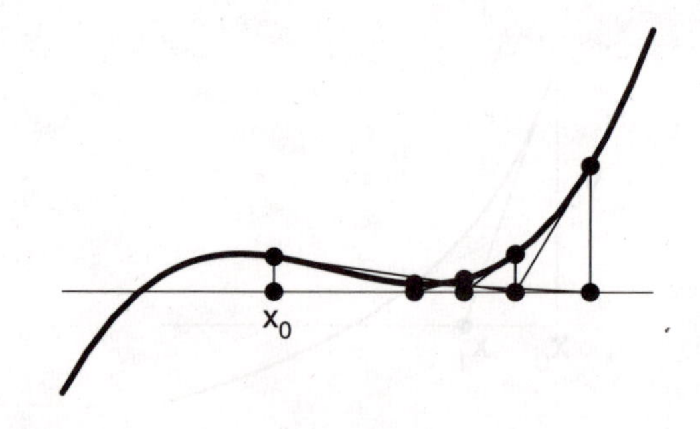

Figure 10.6. A failure of the Newton-Raphson method.

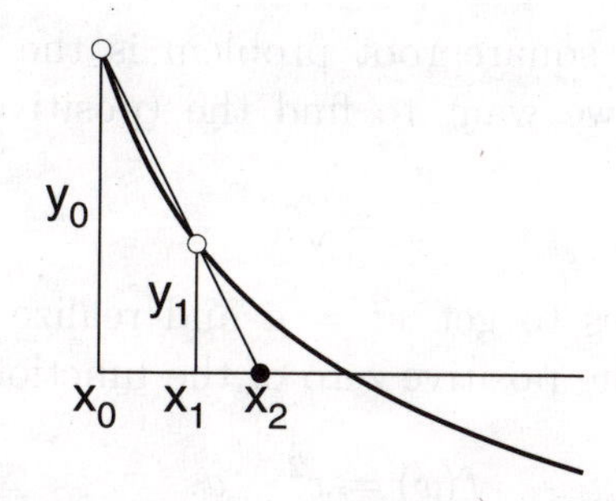

Figure 10.7. The secant method.

Derivatives are not always available. In such cases, the Newton-Raphson method is changed to the *secant method.* We now assume that we have two function values y_0 and y_1 at x_0 and x_1, respectively; see Figure 10.7. We approximate the function $y(x)$ by the secant through the two given points and compute the zero x_2 from

$$\frac{y_0 - y_1}{x_0 - x_1} = \frac{y_0}{x_0 - x_2},$$

thus obtaining

$$x_2 = x_0 - y_0 \frac{x_0 - x_1}{y_0 - y_1}. \tag{10.3}$$

The process is repeated until a solution is found. Note that $(x_0 - x_1)/(y_0 - y_1)$ in (10.3) is an approximation to the factor $1/y_0'$ in (10.2).

10.3 Case Study: Computing the Square Root

In most math classes, you will find that the quadratic equation

$$ax^2 + bx + c = 0$$

has the *exact* solution

$$x = \frac{-b \pm \sqrt{b^2 - 4ac}}{2a}.$$

The problem is, there is no *exact* way to find the square root of a number. Instead, one has to resort to approximations, and the most popular one for the square root problem is the Newton-Raphson iteration. Suppose we want to find the (positive) square root of some number a:

$$x = \sqrt{a}.$$

We square both sides to get $x^2 = a$ and realize that we are now attempting to find the positive zero of the function

$$f(x) = x^2 - a.$$

Next, let's pick a starting value x_0. We observe $f'(x) = 2x$ and obtain the iterative procedure

$$x_{i+1} = x_i - \frac{x_i^2 - a}{2x_i}.$$

This will converge for any positive choice of x_0. For a concrete example, let's take $a = 2$ and $x_0 = 1$. We obtain the sequence

$$\begin{array}{l} 1 \\ 1.5 \\ 1.41667 \\ 1.41422 \\ 1.41421 \\ 1.41421, \end{array}$$

thus quickly approaching the correct value for $\sqrt{2}$.

10.4 Bisection

In a situation where we know that $f(a)$ and $f(b)$ have different signs, we know that there must be at least one zero of f in the interval $[a, b]$. We will create a sequence of intervals that eventually will converge to a root of f. Take the midpoint c of $[a, b]$: it is given by $c = (a+b)/2$. Then $f(c)$ either is zero (within a tolerance) and we have found a zero, or it will have a sign change with either $f(a)$ or with $f(b)$. We take as our new interval $[a, c]$ or $[c, b]$ (the one with the sign change) and continue. At each step, the interval length is halved. After n steps, we have an interval of length $(b-a)/2^n$. If this is less than a preassigned tolerance ϵ_x and the function values differ by less than a tolerance ϵ_y, then the midpoint of the interval is taken to be the root.

This method is very simple and robust. Although it is not very fast, it is guaranteed to find a solution.

10.5 Case Study: Wilkinson Polynomials

During the 1950s and 1960s, computer-based numerical algorithms started to appear. For many of them, scientists encountered problems, which we now understand in terms of roundoff errors and the condition of an algorithm. A famous example of a simple problem that is intractable numerically is due to J. Wilkinson.

Let

$$w(t) = (t-1) \cdot (t-2) \cdot \ldots \cdot (t-n)$$

be a polynomial of degree n with n real zeroes $1, 2, \ldots, n$. It is called *Wilkinson's polynomial.*

For our case study, we presented Mathematica with high-degree ($n = 100$) versions of this polynomial and asked for the zeroes—they were flawlessly found from the factored form. We then expanded $w(t)$ into the monomial form, and again asked to find the zeroes. This time, however, from $n = 23$ on, incorrect complex zeroes were found!

10.6 Problems and Experiments

1. Modify the algorithm in Section 10.3 to find cube roots; that is, for a given x, find $x^{1/3}$.

2. Experiment with the given square root algorithm. Check how many new correct digits you obtain per step. Then implement bisection. How many new correct digits do you get per step?

3. Repeat Problem 2 for the root of $y = x^4$. You should notice a much slower behavior of the Newton-Raphson method, caused

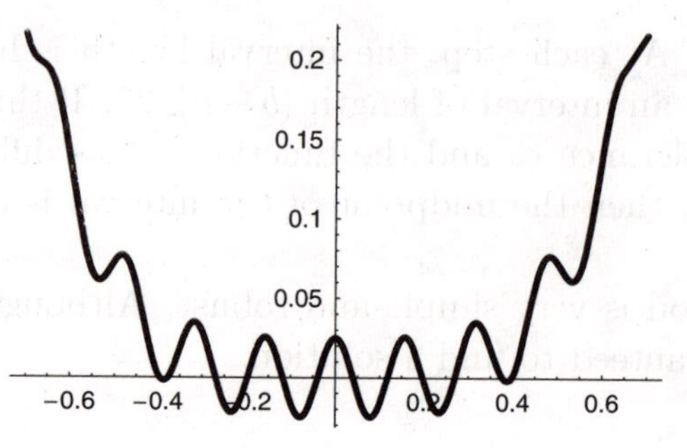

Figure 10.8. A function with several roots.

by the presence of a *multiple root* at $x = 0$. What can you say about the performance of the bisection method?

4. Figure 10.8 shows the function

$$f(x) = x^4 + \cos(40x)/40$$

over the interval $[-0.7, 0.7]$. Use any method you like to find *all* roots in that interval.

11

Computing with Multivariate Functions

In this chapter, we focus on functions of more than one variable. Primarily, the focus is on functions of two and three variables, called bivariate and trivariate functions, respectively. We look at derivatives and integration over these functions, which are useful for volume calculations. Also presented in this chapter are quadratic forms, which are simple but special bivariate functions that appear frequently in science and engineering applications. Continuing the discussion from Section 10.2, we introduce a bivariate Newton-Raphson method for finding intersections. We continue the discussion of PDEs as well, and look at some special bivariate functions defined by them.

11.1 Bivariate Functions

A function assigns a value $f(x)$ to the variable x; it maps an interval of the real axis to the reals. More general types of functions solve more general tasks: they assign function values to 2D points (x, y). An example is a weather map displaying temperature as a function of location (x, y). A more formal example is

$$f(x, y) = \sin(x^2 + y^2).$$

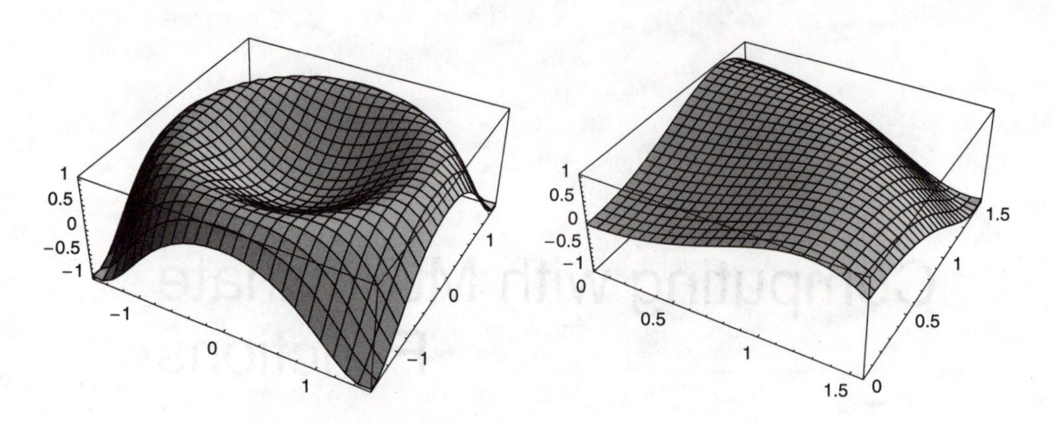

Figure 11.1. Plot of the function $\sin(x^2 + y^2)$ over $-\pi/2 \leq x \leq \pi/2$ and $-\pi/2 \leq y \leq \pi/2$. (left), and over $0 \leq x \leq \pi/2$ and $0 \leq y \leq \pi/2$ (right).

This type of function is known as a *bivariate function* because it has two variables. A display of this function is shown in Figure 11.1, plotted over two different domains.

One of the simplest bivariate functions is a *linear* one of the form

$$l(x, y) = ax + by + c.$$

This kind of function describes a *plane*. For any point (x, y), the plane has its largest slope in the direction (a, b). Figure 11.2 shows

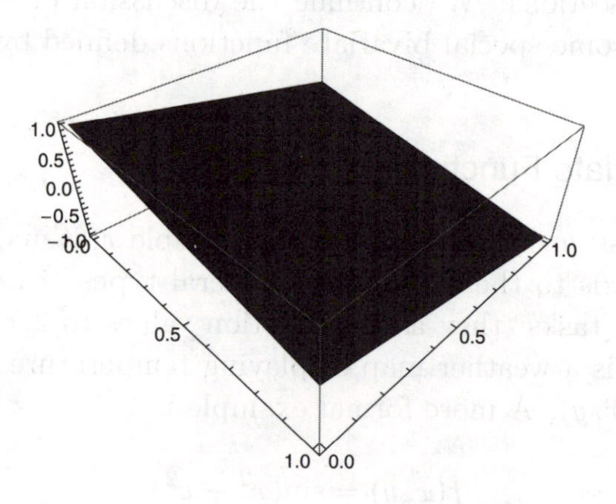

Figure 11.2. The plane $l(x, y) = 1 - x - y$.

the plane $l(x,y) = 1 - x - y$. Its largest slope is in the direction $(-1,-1)$.

We next discuss the concept of derivatives. For simplicity, we assume our function f is defined over the *unit square* $0 \leq x \leq 1$, $0 \leq y \leq 1$. Let $y = c$ be a straight line parallel to the x-axis. If we restrict f to this line, we obtain a univariate function $f(x,c)$, which we can differentiate with respect to x. This gives the tangent to the curve at (x,c), which is shown in Figure 11.3. The slope of $f(x,c)$ is written as

$$\frac{\partial f(x,c)}{\partial x} \quad \text{or} \quad f_x(x,c),$$

which is called a *partial derivative* of $f(x,y)$. In the same manner, we can obtain slopes in the y-direction.

In the univariate case, the tangent vector at $(x, f(x))$ is given by $(1, f'(x))$. In the bivariate case, it makes sense to speak of a *tangent plane* at the point $(x_0, y_0, f(x_0, y_0))$. The tangent plane is a linear function $l(x,y)$ of x and y such that

$$l(x,y) = f(x_0,y_0) + (x - x_0)f_x(x_0,y_0) + (y - y_0)f_y(x_0,y_0). \quad (11.1)$$

This plane has its steepest slope in the direction $(f_x(x_0,y_0), f_y(x_0,y_0))$. Since the tangent plane locally approximates the function f, it follows that f also has its steepest slope in that direction. This leads to the definition of the *gradient* ∇f:

$$\nabla f = (f_x, f_y). \quad (11.2)$$

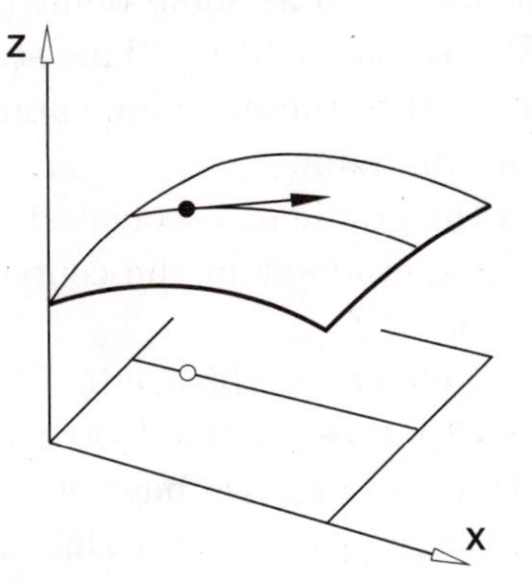

Figure 11.3. Geometry of a partial derivative.

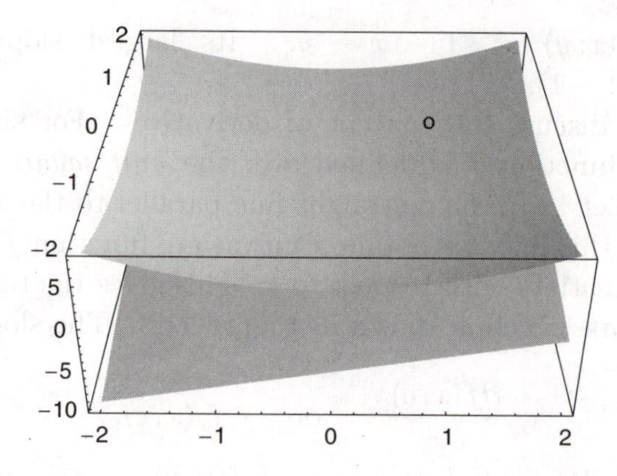

Figure 11.4. A function (upper) and a tangent plane (lower). The point of contact is marked.

We can thus shorten (using $\mathbf{x} = (x, y)$) the tangent plane equation (11.1) to

$$l(\mathbf{x}) = f(\mathbf{x}_0) + \nabla f(\mathbf{x}_0) \cdot (\mathbf{x} - \mathbf{x}_0). \qquad (11.3)$$

Figure 11.4 shows an example of a function and a tangent plane. The function is $f(x, y) = x^2 + y^2$. The tangent plane at $(x_0, y_0) = (1, 1)$ is given by

$$l(x, y) = f(1, 1) + \nabla f(1, 1) \cdot \begin{bmatrix} x - 1 \\ y - 1 \end{bmatrix} = -2 + 2x + 2y.$$

A function has a *minimum* at some point (x_0, y_0) if its gradient is the zero vector: $\nabla f(x_0, y_0) = (0, 0)$. This states that the tangent plane at (x_0, y_0) is parallel to the x, y-plane and thus agrees with our intuitive concept of a minimum.

Figure 11.5 shows the gradients associated with a bivariate function. Note the vanishing gradient in the center, corresponding to a horizontal tangent plane of f.

Integrals are also defined for bivariate functions. We restrict ourselves to functions that are defined over a rectangle $R : a \leq x \leq b, c \leq y \leq d$. Let us assume we have x-values $x_0, \ldots, x_M$ with spacing given by $\Delta x_i = x_{i+1} - x_i$, and that $x_0 = a$ and $x_M = b$. Similarly, assume we have values $y_0, \ldots, y_N$ on the interval $[b, c]$. These two partitions generate a rectangular grid on the rectangle

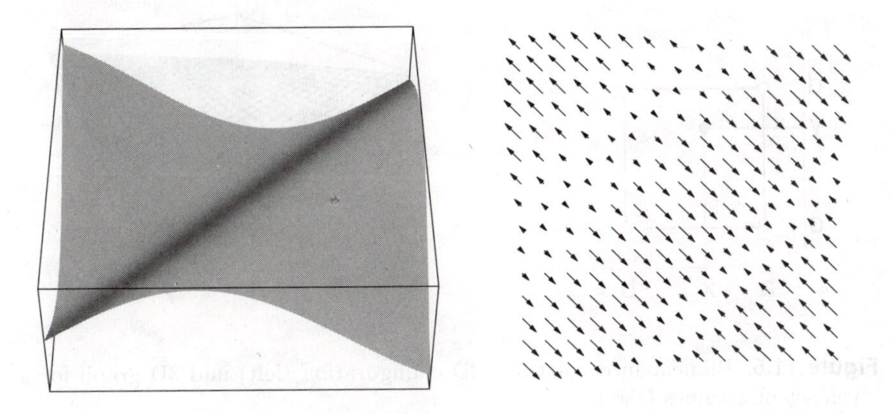

Figure 11.5. The function $f(x, y) = \sin(x - y)$ (left) and its gradient field (right).

R. We associate $f(x_{i+1}, y_{j+1})$ with the grid cell whose side lengths are Δx_i and Δy_j. The volume over this cell is given by area times height: $\Delta x_i \Delta y_j f(x_{i+1}, y_{j+1})$. If we sum over all these volumes, we get an approximation for the volume under the surface:

$$V = \sum_{i=0}^{M-1} \sum_{j=0}^{N-1} \Delta x_i \Delta y_j f(x_{i+1}, y_{j+1}) \tag{11.4}$$

in complete analogy to the area calculation using (7.3).

Again, taking the limit $M, N \to \infty$, the double sum converges to a *double integral*

$$V = \int_a^b \int_c^d f(x, y) \mathrm{d}x \mathrm{d}y. \tag{11.5}$$

11.2 Bilinear Interpolation

In the field of Geographic Information Systems (GIS), satellite data are frequently acquired on a rectangular grid. For a small area (say, the size of most US states), we may safely ignore Earth's curvature and work with a planar rectangular grid. At each 2D grid vertex $\mathbf{x}_{i,j}$, we are given an altitude $z_{i,j}$. For some applications, it may be necessary to find altitudes for points other than the given $\mathbf{x}_{i,j}$. This problem is solved by *bilinear interpolation.*

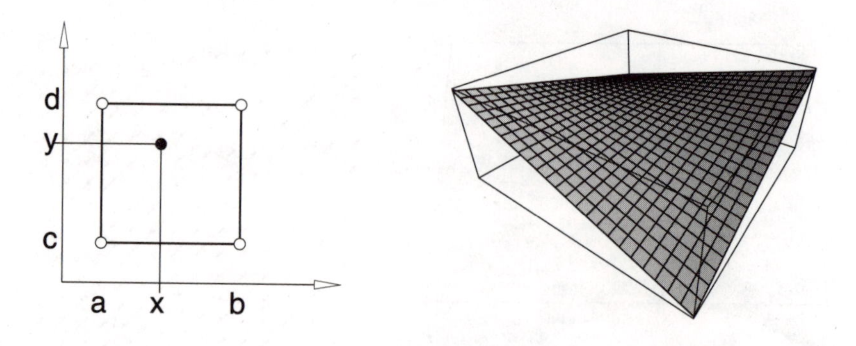

Figure 11.6. Bilinear interpolation: 2D configuration (left) and 3D graph for a given set of z-values (right).

In Section 7.1, we encountered linear interpolation for functions of one variable. Now we need the analogue for functions of two variables. As illustrated in Figure 11.6, we are given four 2D points

$$\begin{bmatrix} (a,c) & (b,c) \\ (a,d) & (b,d) \end{bmatrix} \quad \text{and function values} \quad \begin{bmatrix} z_{a,c} & z_{b,c} \\ z_{a,d} & z_{b,d} \end{bmatrix},$$

and we ask, for a given 2D point (x,y), what is a reasonable estimate for its function value z? The answer is

$$z = \begin{bmatrix} \frac{b-x}{b-a} & \frac{x-a}{b-a} \end{bmatrix} \begin{bmatrix} z_{a,c} & z_{a,d} \\ z_{b,c} & z_{b,d} \end{bmatrix} \begin{bmatrix} \frac{d-y}{d-c} \\ \frac{y-c}{d-c} \end{bmatrix}. \tag{11.6}$$

In order to make sense of this, check that $(x,y) = (a,b)$ yields $z_{a,b}$, and continue for the three remaining points. As a requirement of linear interpolation, the coefficients for interpolation in the x- and y-directions must sum to one. Breaking down (11.6), we see it is equivalent to three linear interpolation steps. For instance, multiplying the first and second matrix results in

$$z_1 = \frac{b-x}{b-a} z_{ac} + \frac{x-a}{b-a} z_{bc} \quad \text{and} \quad z_2 = \frac{b-x}{b-a} z_{ad} + \frac{x-a}{b-a} z_{bd}.$$

Then the final function value is computed as

$$z = \frac{d-y}{d-c} z_1 + \frac{y-c}{d-c} z_2.$$

Alternatively, we could compute two function values in the y-direction first.

Figure 11.7. Three quadratic functions (top), defined (from left to right) by the matrices A_1, A_2, and A_3, respectively, and their contours (bottom).

11.3 Quadratic Forms

An important yet simple class of bivariate functions are the *quadratic forms.* They are of the form

$$q(\mathbf{x}) = \mathbf{x}^{\mathrm{T}} A\mathbf{x} + c \tag{11.7}$$

where $\mathbf{x}^{\mathrm{T}} = [x, y]$ and A is a symmetric matrix. We quickly see that (11.7) indeed is a quadratic function of two variables x and y by rewriting:

$$q(\mathbf{x}) = a_{1,1}x^2 + 2a_{1,2}xy + a_{2,2}y^2 + c.$$

We already encountered such a function in Figure 11.4.

The constant c affects the z-elevation of the function; it does not influence its shape. Shape is determined by the matrix A. Figure 11.7 shows three quadratic functions defined by the matrices

$$A_1 = \begin{bmatrix} 3 & 0.5 \\ 0.5 & 1 \end{bmatrix}, \quad A_2 = \begin{bmatrix} 1 & -2 \\ -2 & 1 \end{bmatrix}, \quad A_3 = \begin{bmatrix} 2 & 0 \\ 0 & 0 \end{bmatrix}.$$

The eigenvalues of the matrices are $3.1, 0.89$ for A_1; $3, -1$ for A_2; and $2, 0$ for A_3. For A_1, we also note that its dominant eigenvector is $[0.97, 0.23]^{\mathrm{T}}$. It is in this direction that the highest rate of change takes place.

The eigenvalues of the matrices are key to understanding the shape of the functions. If all eigenvalues are positive, then A is called *positive definite.* Quadratic forms generated by positive definite matrices always have exactly one minimum. Thus the quadratic form corresponding to A_1 has one minimum, the one corresponding to A_2 has none, and the one corresponding to A_3 has infinitely many minima.

Another important aspect of quadratic forms is the shape of the surface defined by (11.7). We see that A_1 produces an elliptic paraboloid, A_2 produces a hyperbolic paraboloid, and A_3 produces a cylindrical paraboloid.

Quadratic forms are not limited to 2×2 matrices, however. Indeed, (11.7) defines a quadratic form for any number of variables[1] if A is a symmetric $n \times n$ matrix. If A has positive eigenvalues, then the quadratic form has exactly one minimum.

11.4 Contouring

Bivariate functions have zeroes just as univariate functions do. The difference is that the zeroes of univariate functions are distinct points, whereas bivariate functions have *curves* as their zero sets. These curves are the intersections with the plane $z = 0$ and the graph $z = f(x, y)$ of the function f. Often one is interested in the intersections with other planes $z = c$ as well, resulting in curves that are called *isolines* or *contours.* More precisely, if we have a bivariate

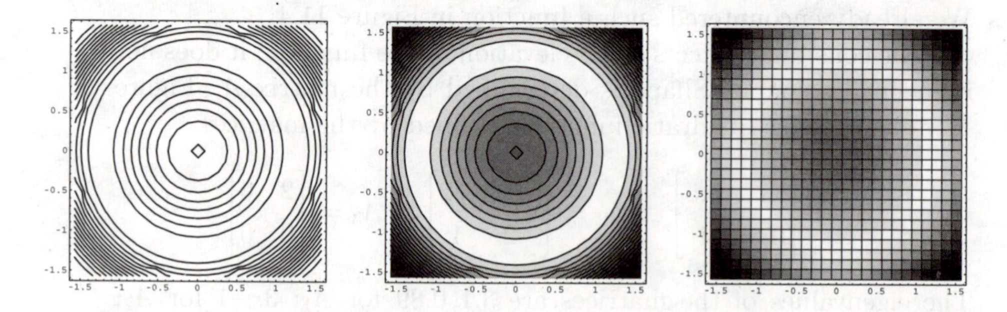

Figure 11.8. Contour plot styles: contours (left), shaded contours (middle), and density plot (right).

[1]That is, $\mathbf{x} = [x_1, \ldots, x_n]^{\mathrm{T}}$.

function $f(x, y)$, then an isoline is the set of all points (x, y) that satisfy the equation

$$f(x, y) = c \tag{11.8}$$

where the constant c defines the desired isolevel. This equation does not necessarily describe just one curve; there can be many!

Figure 11.8 shows a set of different contours of the function of Figure 11.1. The middle and right contours contain shading information: white corresponds to large function values, and black corresponds to low function values. The plots were generated with Mathematica's `ContourPlot` function. Looking carefully near the borders, we see that the contours appear to be wrong—they all should be exact circles! For now, let's accept this; we explore this problem further and just how bivariate contours are computed in Section 14.3.

The contours of two bivariate functions $f(x, y)$ and $g(x, y)$ are shown in Figure 11.9. In the x, y-plane, the intersection of the functions is given by all points (x, y) with $f(x, y) - g(x, y) = 0$, as illustrated in the right part of the figure.

We will continue to explore this valuable tool. A case study for contouring is provided in Section 14.4. Contours of trivariate functions are explored briefly in Section 11.7, and then in more detail in Section 15.2.

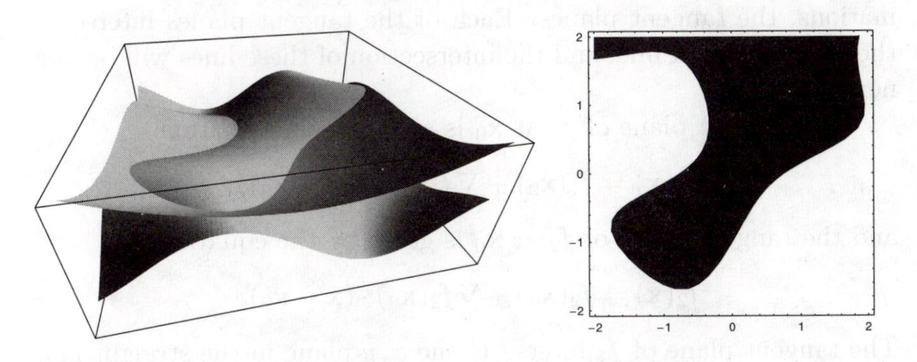

Figure 11.9. Two intersecting bivariate functions (left) and the zero contour of the difference between these functions (right).

11.5 The Newton-Raphson Method

Let's consider two functions $f_1(x, y)$ and $f_2(x, y)$. The zero ($c = 0$) contour of $f_1(x, y)$ is a curve in the x, y-plane, and the zero contour of $f_2(x, y)$ is also a curve in the x, y-plane. At points (x_i, y_i) where both curves intersect, both f_1 and f_2 are zero. Finding these common zeroes of two functions is typically formulated as

$$\begin{bmatrix} f_1(x, y) \\ f_2(x, y) \end{bmatrix} = \begin{bmatrix} 0 \\ 0 \end{bmatrix}$$

or, more concisely, by

$$\mathbf{f}(\mathbf{x}) = \mathbf{0}.$$

For example, let's consider

$$\mathbf{f}(\mathbf{x}) = \begin{bmatrix} x^2 + y^2 - 1 \\ xy - 0.5 \end{bmatrix} = \begin{bmatrix} 0 \\ 0 \end{bmatrix} = \mathbf{0}.$$

The first equation, $f_1(x, y) = x^2 + y^2 - 1$, describes a paraboloid. The second equation, $f_2(x, y) = x + y$, describes a hyperboloid. The zero contour of f_1 is a circle, the zero contour of f_2 is a hyperbola. Where do these two curves intersect? This problem is solved by the bivariate Newton-Raphson method. (The univariate Newton-Raphson was discussed in Section 10.2.) Suppose we have a guess of $\mathbf{x}_0$ for such an intersection point. This point will give us two function values, namely $f_1(\mathbf{x}_0)$ and $f_2(\mathbf{x}_0)$. In the spirit of the Newton-Raphson method, we will replace the functions f_1 and f_2 by linear approximations, the tangent planes. Each of the tangent planes intersects the x, y-plane in a line, and the intersection of these lines will be our next guess, $\mathbf{x}_1$.

The tangent plane of f_1 at $\mathbf{x}_0$ is given by the equation

$$l_1(\mathbf{x}) = f_1(\mathbf{x}_0) + \nabla f_1(\mathbf{x}_0) \cdot (\mathbf{x} - \mathbf{x}_0),$$

and the tangent plane of f_2 at $\mathbf{x}_0$ is given by the equation

$$l_2(\mathbf{x}) = f_2(\mathbf{x}_0) + \nabla f_2(\mathbf{x}_0) \cdot (\mathbf{x} - \mathbf{x}_0).$$

The tangent plane of f_1 intersects the x, y-plane in the straight line $l_1(\mathbf{x}) = 0$, and that of f_2 in the straight line $l_2(\mathbf{x}) = 0$. The intersection of these two straight lines yields our next point, $\mathbf{x}_1$. By combining the two equations above into a more concise notation, we get

$$\mathbf{0} = \mathbf{f}(\mathbf{x}_0) + \begin{bmatrix} \nabla f_1(\mathbf{x}_0) \\ \nabla f_2(\mathbf{x}_0) \end{bmatrix} \cdot (\mathbf{x}_1 - \mathbf{x}_0),$$

and striving for even more conciseness, we define

$$J(\mathbf{x}_0) = \begin{bmatrix} \nabla f_1(\mathbf{x}_0) \\ \nabla f_2(\mathbf{x}_0) \end{bmatrix} = \begin{bmatrix} \frac{\partial f_1(\mathbf{x}_0)}{\partial x} & \frac{\partial f_1(\mathbf{x}_0)}{\partial y} \\ \frac{\partial f_2(\mathbf{x}_0)}{\partial x} & \frac{\partial f_2(\mathbf{x}_0)}{\partial y} \end{bmatrix}. \tag{11.9}$$

The 2×2 matrix $J(\mathbf{x}_0)$ is the *Jacobian* matrix of $\mathbf{f}$ at $\mathbf{x}_0$. We now have

$$\mathbf{0} = \mathbf{f}(\mathbf{x}_0) + J(\mathbf{x}_0) \cdot (\mathbf{x}_1 - \mathbf{x}_0),$$

and solve for $\mathbf{x}_1$:

$$\mathbf{x}_1 = \mathbf{x}_0 - J^{-1}(\mathbf{x}_0) \cdot \mathbf{f}(\mathbf{x}_0). \tag{11.10}$$

We must keep in mind that this "solution" may not exist: as an example, we consider the case of both tangent lines $l_1(\mathbf{x}_0)$ and $l_2(\mathbf{x}_0)$ being parallel. This would mean that the two gradient vectors $\nabla f_1(\mathbf{x}_0)$ and $\nabla f_2(\mathbf{x}_0)$ are parallel. Then there is no intersection. But as long as an intersection exists, and if $\mathbf{x}_0$ is close to the final solution, we may repeat (11.10) as

$$\mathbf{x}_{i+1} = \mathbf{x}_i - J^{-1}(\mathbf{x}_i) \cdot \mathbf{f}(\mathbf{x}_i) \tag{11.11}$$

until we converge to a solution $\mathbf{x}$. This process is illustrated in Figure 11.10.

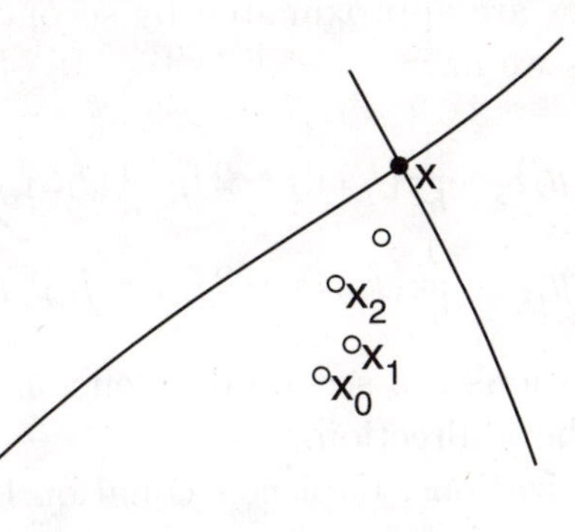

Figure 11.10. An example of a few Newton-Raphson steps.

11.6 Partial Differential Equations

Let's consider the following scenario. We have four metal wires forming four curves, as shown in Figure 11.11 (left). We are trying to model a soap bubble that would be created by immersing the wires in a soap solution and pulling them out. A reasonable outcome would look like the surface shown in Figure 11.11 (right). Let us investigate that figure a little, trying to find the method to produce it. The surface in Figure 11.11 represents a bivariate function $f(x, y)$. At any point, it is convex up in one direction and convex down in the other. This means that at every point, f_{xx} and f_{yy} have opposite signs. In fact, for our problem, they need to be equal in magnitude, leading to

$$f_{xx}(x, y) + f_{yy}(x, y) = 0. \tag{11.12}$$

This is a *partial differential equation*—partial because it involves partial derivatives—and it is called *Laplace's equation.*

Let's assume our function is defined over the unit square $0 \leq x, y \leq 1$. Our input data are the four boundary curves; they are

$$f(0, y), \quad f(x, 0), \quad f(1, y), \quad f(x, 1).$$

These input functions are called *boundary conditions.*

In order to solve the Laplace problem, we put an $(n+1) \times (n+1)$ grid of points (x_i, y_j) on the unit square. The number n is referred to as the *resolution* of the grid. We assume uniform spacing h in both the x- and the y-direction. We abbreviate $f_{i,j} = f(x_i, y_j)$.

We now need to discretize (11.12). For this you should recall that second derivatives are approximated by second differences (see Section 8.8), and thus we have

$$f_{xx}(x_i, y_j) \approx \frac{1}{h^2}[f_{i+1,j} - 2f_{i,j} + f_{i-1,j}],$$
$$f_{yy}(x_i, y_j) \approx \frac{1}{h^2}[f_{i,j+1} - 2f_{i,j} + f_{i,j-1}].$$

The first of these equations is a second difference in the x-direction, the second one is in the y-direction.

Using these discretizations, Laplace's equation becomes

$$f_{i+1,j} - 2f_{i,j} + f_{i-1,j} + f_{i,j+1} - 2f_{i,j} + f_{i,j-1} = 0.$$

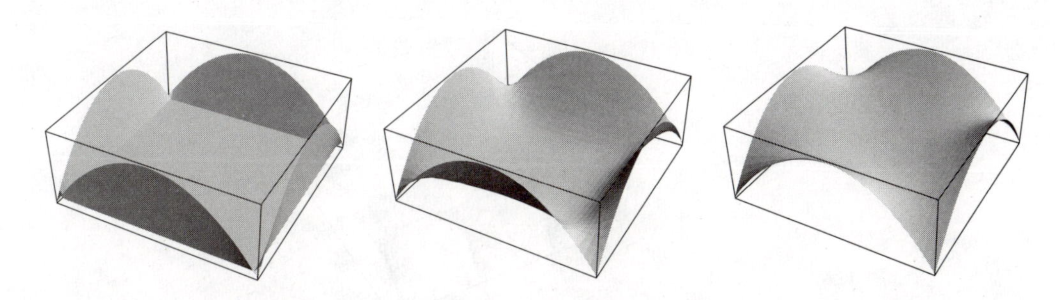

Figure 11.11. Solving Laplace's equation: input data (left), solution after 20 iterations (middle), and solution after 150 iterations (right).

Solving for $f_{i,j}$ yields

$$f_{i,j} = [f_{i+1,j} + f_{i-1,j} + f_{i,j+1} + f_{i,j-1}]/4. \tag{11.13}$$

This is more conveniently written using an *update mask*:

$$f_{i,j} \quad = \quad \begin{matrix} 0 & 0.25 & 0 \\ 0.25 & \bullet & 0.25. \\ 0 & 0.25 & 0 \end{matrix} \tag{11.14}$$

We now have a way to solve the discrete Laplace problem. Start by assigning values $f_{i,j} = 0$ except for those on the four boundaries—this is the situation depicted in Figure 11.11 (left). Then, looping over all $1 \le i, j \le n-1$, we use (11.13) as an *update* for $f_{i,j}$. Having finished one update cycle, we continue with more. An intermediate result of this iterative process is also shown in Figure 11.11 (middle). Repeating sufficiently many times will converge to the final solution, again in Figure 11.11 (right).

Some discussion seems in order here. Our update process was really nothing other than solving a linear system of equations by using the Gauss-Seidel iteration, which was introduced in Section 5.6. There are $(n-2) \times (n-2)$ equations of the form (11.13). These equations linearly involve the $(n-2) \times (n-2)$ unknowns as well as the known boundary values. Thus we have a linear system with $(n-2) \times (n-2)$ equations in as many unknowns. Instead of solving it by using Gauss elimination, the iterative Gauss-Seidel method is a more natural choice.

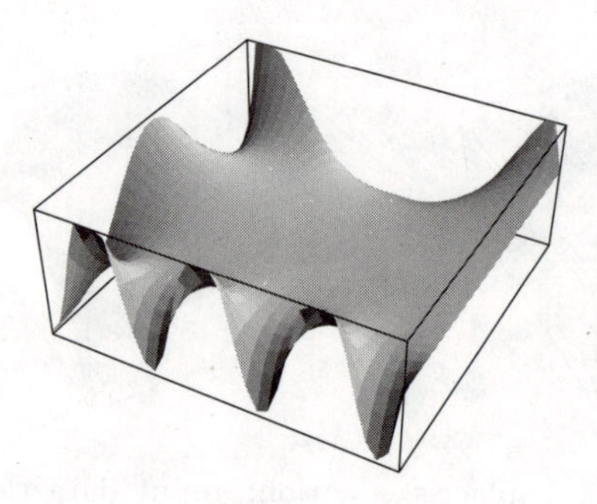

Figure 11.12. A different Laplace problem.

We have not touched upon some practical issues: What should the grid spacing h be? How many update cycles do we need?[2] These are questions that often depend on the application at hand and will not be covered here.

Let's consider another example of the discrete Laplace problem. The boundary curves in our examples so far have all all been identical; however, this is not generally so. Figure 11.12 gives a different example.

PDEs are not limited to the case of Laplace's equation. Any relationship between partial derivatives of a bivariate function plus some known data of the function qualifies as a PDE. We give one more example, the *Euler-Lagrange PDE*, given by

$$f_{xxyy} = 0 \tag{11.15}$$

together with boundary conditions just as for Laplace's equation. Again, an iterative solution is obtained by discretizing (11.15). For each 3×3 subgrid, we take second differences for all three rows and then the second differences of the resulting three values. This leads to the following update mask; the solution is again obtained iteratively:

$$f_{i,j} = \begin{matrix} -0.25 & 0.50 & -0.25 \\ 0.50 & \bullet & 0.50 \\ -0.25 & 0.50 & -0.25 \end{matrix} . \tag{11.16}$$

[2]For the example in Figure 11.11, we used $h = 0.03$ and 150 iterations for the final answer.

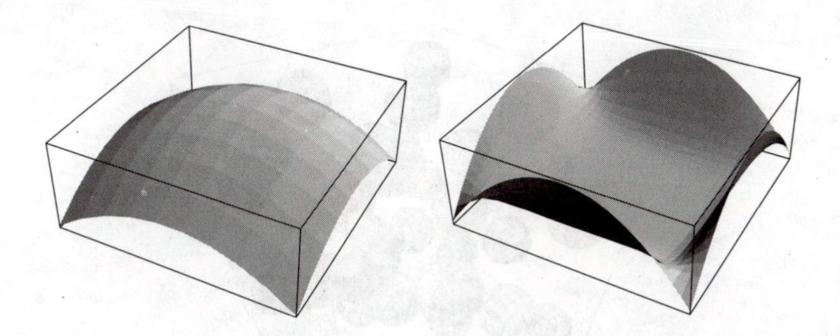

Figure 11.13. Solving the Euler-Lagrange equation: a very close approximation to the true solution after 600 iterations over a 10×10 grid (left), and the approximation after 600 iterations using a 30×30 grid (right).

Figure 11.13 gives an example. The boundary conditions are the same as in Figure 11.11. Figure 11.13 illustrates that speed of convergence is an issue. In the left part of the figure, we used 600 iterations over a 10×10 grid and obtained a very good approximation to the true solution. In the right part, we see that 600 iterations are totally inadequate for a 30×30 grid! As a rule, a preset number of iterations is not a good idea. We should instead employ a stopping criterion, such as stopping when the change between two successive iterations is below a preset tolerance. This applies to all iterative methods, not just this specific example.

11.7 Trivariate Functions

A *trivariate function* is of the form

$$w = f(x, y, z). \tag{11.17}$$

Trivariate functions cannot easily be visualized with plots such as in Figure 11.1 since their graphs "live" in four dimensions (4D). To better understand the nature of these functions, let's look at an example. A room in a building can be equipped with a coordinate system such that each point in it is assigned a coordinate triple (x, y, z). Let's further assume we can measure the temperature t at each point. Then $t(x, y, z)$ is the trivariate function describing the temperature distribution in the room.

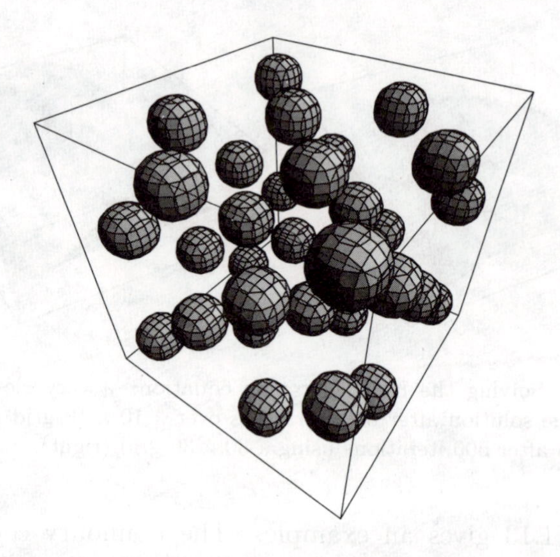

Figure 11.14. Example of a contour plot of a trivariate function.

Finding all points in the room where the temperature has a constant value c leads to the problem of *contouring*. A contour of a trivariate function is a set of bivariate surfaces, given by $f(x, y, z) = c$. For example, $w = x^2 + y^2 + z^2$ describes a 4D paraboloid. Its contour $x^2 + y^2 + z^2 = c$ describes a sphere, centered at the origin. Finding contours of general trivariate functions is a tricky problem; see Section 15.2 for a solution.

A contouring example is shown in Figure 11.14. The trivariate function is

$$w = \sin 8x \sin 8y \sin 8z.$$

The contour value is $w = 0.45$. Try to make sense of the complex pattern!

Trivariate functions are differentiated in total analogy to bivariate ones. For example, the gradient ∇ is given by

$$\nabla f(x, y, z) = (f_x, f_y, f_z).$$

For any point (x, y, z), it represents the vector pointing in the direction in which f changes the most. For the example $w = x^2 + y^2 + z^2$, we have

$$\nabla w = (2x, 2y, 2z).$$

An important trivariate function is the *trilinear interpolant.* It takes eight scalar values, given at the vertices of a cube, and finds a scalar value at any location inside the cube. Denote the cube vertices by $\mathbf{v}_{0,0,0} = (a,b,c)$, $\mathbf{v}_{1,0,0} = (a,b,d), \ldots, \mathbf{v}_{1,1,1} = (d,e,f)$, and the corresponding scalar values by $w_{i,j,k}$. Then a point $\mathbf{x} = (x,y,z)$ inside the cube is assigned a function value in the following way. This is a straightforward generalization of the bilinear case (11.6), but let's simplify the notation by assigning

$$r = \frac{x-a}{b-a} \qquad 1-r = \frac{b-x}{b-a}$$
$$s = \frac{y-c}{d-c} \qquad 1-s = \frac{d-y}{d-c}$$
$$t = \frac{z-e}{f-e} \qquad 1-t = \frac{f-z}{f-e}.$$

Then

$$w(\mathbf{x}) = \sum_{i=0}^{1}\sum_{j=0}^{1}\sum_{k=0}^{1} w_{i,j,k}\left(r^i(1-r)^{1-i}\right)\left(s^j(1-s)^{1-j}\right)\left(t^k(1-t)^{1-k}\right). \tag{11.18}$$

11.8 Problems and Experiments

Several problems will use this function:

$$f(x,y) = 2\sin(6xy) + x^2y^2 + 2x^3.$$

It is plotted in Figure 11.15.

1. What range for x and y is used in Figure 11.15? What viewpoint was used?
2. Experiment with plotting f for various x- and y-ranges.
3. Experiment with changes to f. Try to produce interesting examples.
4. Find the contour $f(x,y) = 0$.

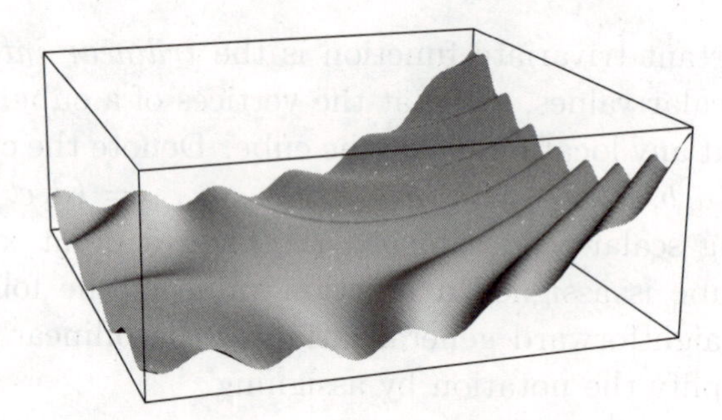

Figure 11.15. A bivariate function.

5. Produce four boundary curves by restricting f to the edges of the square $-2 \leq x, y \leq 2$. Solve Laplace's equation for these boundary curves. Compare your solution to the graph of f.

6. Graphically explore the function

$$g(x, y) = \frac{x}{x + y}; \qquad -1 \leq x, y \leq 1.$$

Take care near the origin; the function has a singularity there.

12

Visualizing Empirical Data

In this chapter, we look at methods for the visualization of one- and two-dimensional empirical data. Here we only scratch the surface of the many variations of visualization tools available. Within a particular scientific discipline, one or more software packages exist that offer widely used methods for that discipline.

12.1 Scatter Plots, Correlations, and Regression

Suppose we are given n values x_i, arising from some observations, and also a set of n values y_i. For example, the x_i may come from measuring the weights of n women and the y_i might measure their heights. A different scenario would be that x_i and y_i are observations from time-dependent experiments; then the sequences (t_i, x_i) and (t_i, y_i) are called *time series*. For example, the x_i might be wind speed and the y_i might be temperature at given times during the course of a day. Clearly it makes most sense to order the data in a times series in the order in which they occurred. In many scientific areas, one likes to know whether there is a linear relationship between the two sets of data. In addition, a visual display of the relation is often desired.

Let's use the following example to demonstrate the concepts to come. Suppose that we have gathered height and weight data from a group of eight women.

	1	2	3	4	5	6	7	8
weight (kg)	65	60	50	52	75	52	72	63
height (cm)	172	170	150	156	172	160	161	160

We set $\mathbf{x} = [x_1, \ldots, x_n]^{\mathrm{T}}$ and $\mathbf{y} = [y_1, \ldots, y_n]^{\mathrm{T}}$, thus creating two n-dimensional vectors $\mathbf{x}$ and $\mathbf{y}$. We further assume that both vectors have their centroid (mean) at the origin:

$$\frac{1}{n}\sum_{i=1}^{n} x_i = 0 \quad \text{and} \quad \frac{1}{n}\sum_{i=1}^{n} y_i = 0. \tag{12.1}$$

For our example, for which $n = 8$, we find that the mean weight is 61.125 kg and the mean height is 162.625 cm. Translating the data by their respective means, the weight vector, $\mathbf{x}$, and the height vector, $\mathbf{y}$, become

$$\mathbf{x} = [3.875, -1.125, -11.125, -9.125, 13.875, -9.125, 10.875, 1.875]^{\mathrm{T}},$$
$$\mathbf{y} = [9.375, 7.375, -12.625, -6.625, 9.375, -2.625, -1.625, -2.625]^{\mathrm{T}}.$$

These vectors are illustrated as data plots in Figure 12.1.

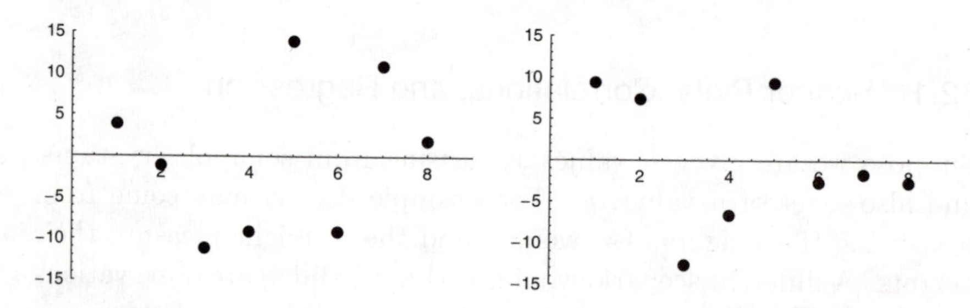

Figure 12.1. Data plots: for weights (i, x_i) (left) and heights (i, y_i) (right)

Another way to visualize the n-dimensional vectors $\mathbf{x}$ and $\mathbf{y}$ is to plot corresponding pairs $\mathbf{r}_i = [x_i, y_i]^{\mathrm{T}}$, resulting in a *scatter plot*. An example is given in Figure 12.2.

The lengths $\|\mathbf{x}\|$ and $\|\mathbf{y}\|$ of our vectors tell us how much they deviate from being the zero vector. The corresponding statistical quantity is called the *standard deviation* σ:

$$\sigma(\mathbf{x}) = \sqrt{\frac{1}{n}(x_1^2 + \ldots + x_n^2)}. \tag{12.2}$$

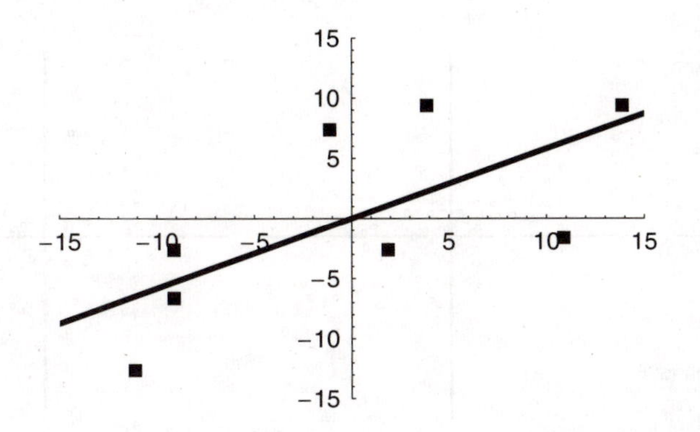

Figure 12.2. The scatter plot of $\mathbf{r}_i = [x_i, y_i]^{\mathrm{T}}$ corresponding to the weight and height data. The line of regression is also shown.

Recall from (12.1) that the origin is the mean of the data, and therefore (12.2) measures the spread of the data about the mean. For $\mathbf{y}$, we compute $\sigma(\mathbf{y})$ in complete analogy. The standard deviation for the weight data set is 9.41 and for the height data set it is 8.01.

We now say the two vectors *correlate* if they point in the same (or in a similar) direction. This is measured by the cosine of the angle between them, which is bounded between -1 and 1. If the cosine is 1, they point in the same direction. If it is zero, they are perpendicular to each other. If it is -1, they point in opposite directions; see Figure 12.3. Hence we use the term *correlation coefficient* ρ for this cosine:

$$\rho = \cos(\mathbf{x}, \mathbf{y}) = \frac{\mathbf{x}^{\mathrm{T}}\mathbf{y}}{\|\mathbf{x}\|\|\mathbf{y}\|}. \tag{12.3}$$

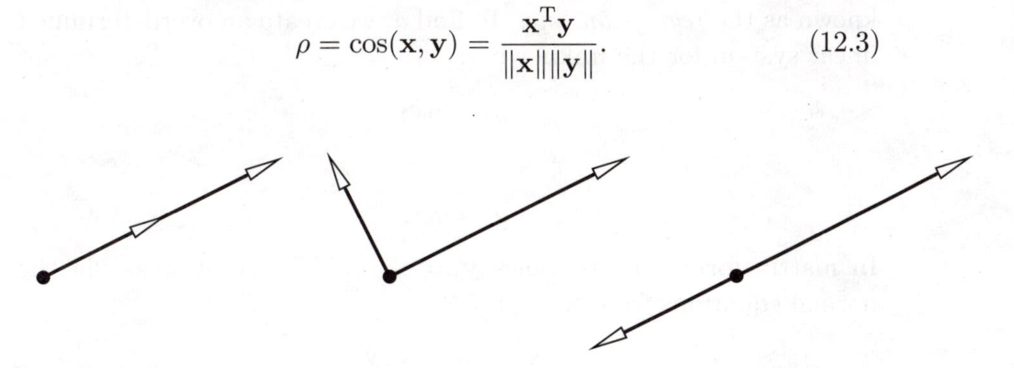

Figure 12.3. Examples of complete correlation ($\rho = 1$) (left), no correlation ($\rho = 0$) (middle), and inverse correlation ($\rho = -1$) (right).

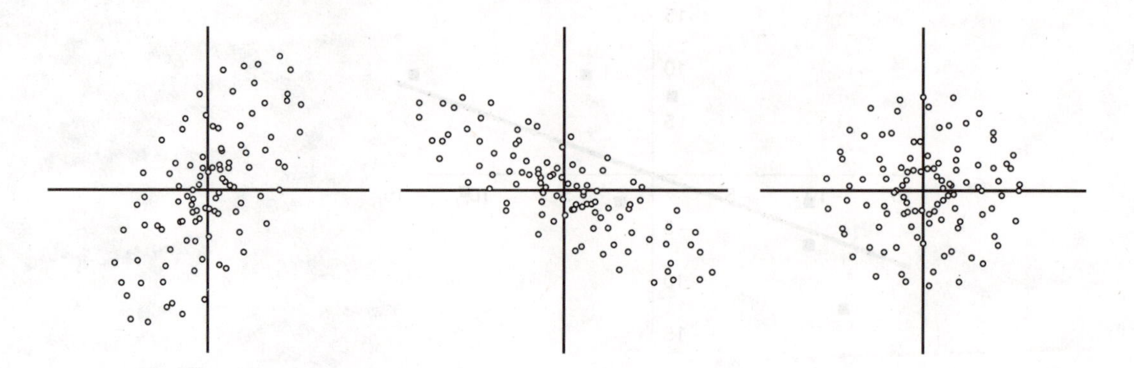

Figure 12.4. Scatter plots and correlation coefficients. From left to right: $\rho = 0.7$, $\rho = -0.8$, $\rho = 0$.

Figure 12.4 shows the three basic types of correlations for some generic data sets. The correlation coefficient for our example is computed as $\rho = 0.68$, indicating that the weights and heights are fairly correlated.

An important property of the correlation coefficient is its *scale invariance*: if both the x_i and the y_i are scaled by the same factor s, then ρ remains the same since both numerator and denominator in (12.3) are scaled the same way.

If the correlation coefficient of the two vectors is large[1] (in absolute value), then we can expect the points $\mathbf{r}_i$ to exhibit a dominant direction. Since the $\mathbf{r}_i$ are centered around the origin, this direction may be described by a straight line of the form $y = ax$. This line is known as the *regression line*. To find a, we create an overdetermined linear system for the unknown a:

$$y_1 = ax_1$$

$$\vdots$$

$$y_n = ax_n.$$

In matrix form, this becomes $\mathbf{y} = \mathbf{x} \cdot a$. The solution is via the normal equations from Section 5.8:

$$\mathbf{x}^{\mathrm{T}}\mathbf{x} \cdot a = \mathbf{x}^{\mathrm{T}}\mathbf{y}.$$

[1]No one definition of a "large" correlation exists, but some practitioners would categorize large as $|\rho| > 0.5$.

Since the matrix $\mathbf{x}^{\mathrm{T}}\mathbf{x}$ is simply a scalar, the solution is obtained directly, without solving a linear system:

$$a = \frac{\mathbf{x}^{\mathrm{T}}\mathbf{y}}{\mathbf{x}^{\mathrm{T}}\mathbf{x}}. \tag{12.4}$$

Note that a always has the same sign as ρ, but not the same magnitude. Also, just as ρ, a is invariant under scalings.

The line of regression for our example is shown in Figure 12.2. Recall that the correlation coefficient, $\rho = 0.68$, indicated that the weights and heights are fairly correlated. This is visually reflected by the positive slope of the line of regression, having slope $a = 0.58$.

We already encountered the regression line in Section 8.4, just not by that name. In the context of that section, we are looking for a linear ($n = 1$) polynomial least squares approximation to the data values. The linear polynomial $y = ax$ is both the regression line and the linear least squares fit.

12.2 PCA Revisited

As in the previous section, assume we are given a point set described by x-coordinates $\mathbf{x} = [x_1, \ldots, x_n]^{\mathrm{T}}$ and y-coordinates $\mathbf{y} = [y_1, \ldots, y_n]^{\mathrm{T}}$. Again, assume the average of each is 0. Both (column) vectors $\mathbf{x}$ and $\mathbf{y}$ may be combined into one matrix R, having n rows and two columns. Being interested in the "shape" of R, we employ the PCA from Section 6.9. That method finds the eigenvalues and eigenvectors of the symmetric 2×2 matrix $R^{\mathrm{T}}R$. The eigenvector corresponding to the largest eigenvalue represents the *dominant line* in our data set.

Let us revisit the example from Section 12.1. The matrix R^{T} is given by

$$R^{\mathrm{T}} = \begin{bmatrix} 3.875 & -1.125 & -11.125 & -9.125 & 13.875 & -9.125 & 10.875 & 1.875 \\ 9.375 & 7.375 & -12.625 & -6.625 & 9.375 & -2.625 & -1.625 & -2.625 \end{bmatrix}.$$

We next determine that

$$R^{\mathrm{T}}R = \begin{bmatrix} 620.875 & 360.375 \\ 360.375 & 449.875 \end{bmatrix},$$

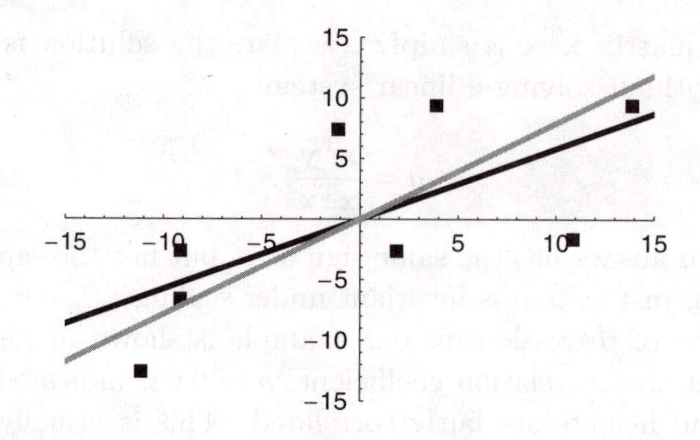

Figure 12.5. The scatter plot corresponding to the weight and height data. The line of regression is black and the dominant line is gray.

has eigenvalues 905.75 and 164.99. The eigenvector corresponding to 905.75 has components 0.78 and 0.62. Thus the dominant line is given by

$$y = \frac{0.62}{0.78}x = 0.79x.$$

The regression line for this example, however, was $y = 0.58x$. Figure 12.5 illustrates both lines. This difference cannot be explained by numerical error—so what is happening? In fact, the regression line and the dominant line solve two different tasks. Take a look at Figure 12.6. On the left, you see a regression line fit. The regression

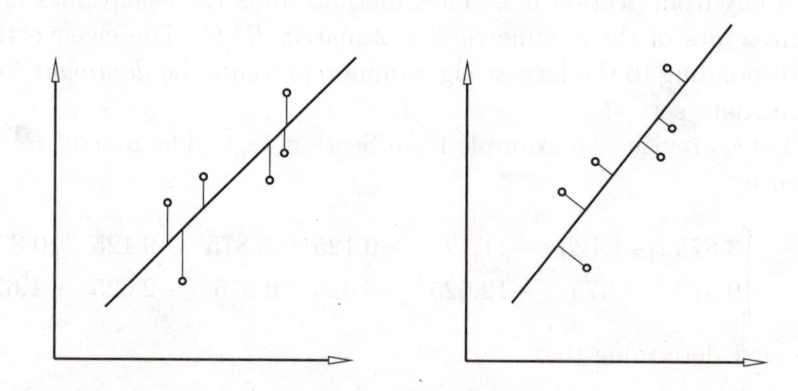

Figure 12.6. Regression line (left) and PCA result (right) for the same data points.

line is computed such that the *vertical* distances from the data points to the regression line are minimized. On the right, you see the PCA result. Here, the *perpendicular* distances to the dominant line are minimized.

An additional way to see the difference between the two methods is by considering a data set in which all $x_i = 0$. The PCA has no problem with this, but the regression line is undefined since it would have infinite slope: the denominator in (12.4) would be zero.

12.3 Histograms, Bar Charts, and Pie Charts

Scientists have to deal with images on a regular basis from sources as diverse as positron emission tomography (PET), GIS, or microscopy. To compare two images with similar content, it is often important to extract salient features and compare them. A standard tool for doing this is the *histogram*, which is a graphical method to visualize frequency distribution; see Section 7.3.

An image is an array of pixels, typically in the range of 500×500 to 2000×2000. Each pixel has three color values for red, green, and blue. For simplicity, here, though, we will deal only with gray-scale images; then each pixel has just one value, ranging from 0 (black) to 1 (white). See Section 16.1 for converting color to gray.

Let us now divide the gray range 0–1 into a number of *bins* or *intervals*, say, 1000, such that for example, the gray level 0.5511 would reside in bin 551. We now ask: for any bin, how many pixels

Figure 12.7. A properly exposed image (left) and its corresponding histogram (right). The histogram is scaled to a maximum bin size of 14,585 pixels.

Figure 12.8. An overexposed image (left) and its corresponding histogram (right). The histogram is scaled to a maximum bin size of 31,499 pixels.

have a gray value that places it in this bin? If we answer this question for every bin, we have a histogram. This is the plot for which the bins are places along the horizontal axis and the number of elements (or frequency) of each bin is plotted on the vertical axis.

Figure 12.7 shows an image on the left and the corresponding gray-scale histogram on the right. The exposure is fairly good, resulting in a somewhat even distribution of gray values. In contrast, Figure 12.8 shows an overexposed image on the left, and the corresponding histogram on the right. Overexposure means more pixels will have higher (lighter) gray values, resulting in a clear shift of the histogram values to the right. Similarly, Figure 12.9 shows an underexposed image on the left, and the corresponding histogram on the right. Now the histogram is shifted to the left, meaning more pixels have low (dark) gray values.

Figure 12.9. An underexposed image (left) and its corresponding histogram (right). The histogram is scaled to a maximum bin size of 85,837 pixels.

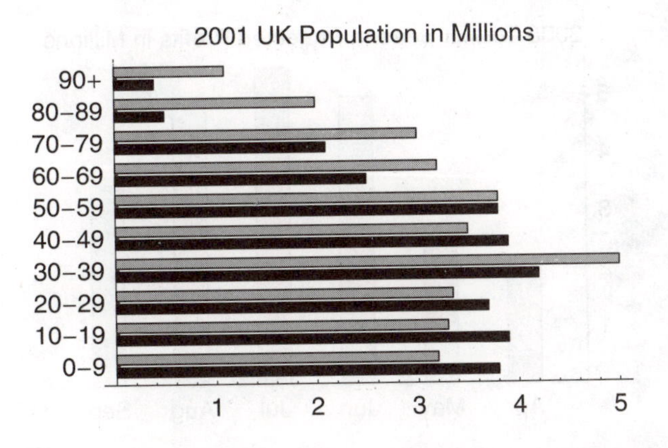

Figure 12.10. Histogram of age distribution of men (black) and women (gray) in the United Kingdom in 2001.

A flaw in the histogram visualization in Figures 12.7 through 12.9 deserves comment. Each histogram has been scaled in the vertical direction based on the maximum bin size of that particular histogram. (These values are given in the captions.) This makes a detailed comparison of the histograms difficult, although we can still see tendencies to lightness or darkness. A better choice would have been to illustrate the histograms over the same scale, which would be determined by the picture with the overall maximum bin size.

Histograms are not limited to images. Figure 12.10 shows a histogram displaying the age distribution of the male and female United Kingdom population in 2001. There are only ten bins, and the frequencies (how many men and women fall into a bin's age group) are plotted horizontally. The choice of the number of intervals can greatly influence the information communicated. This histogram illustrates that more than one data set can be visualized simultaneously, given that the same set of intervals is relevant for both.

A *bar chart* is a visualization tool for categorical (nominal or ordinal) data. Figure 12.11 illustrates an example in which the categories are months. Visually, a bar chart appears identical to a histogram. The difference between the two is that histograms work with interval data which can be discrete or continuous, as

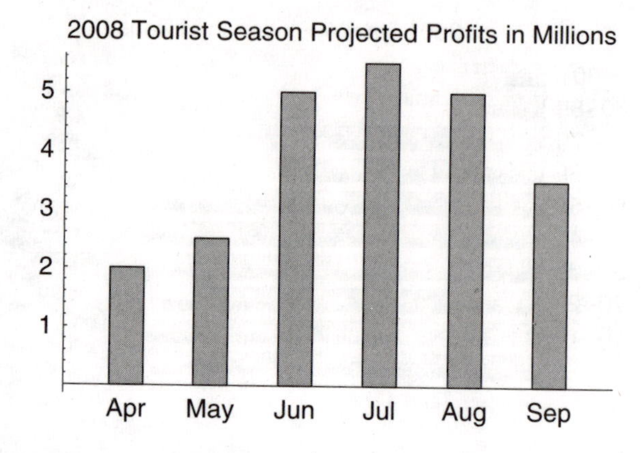

Figure 12.11. Bar chart of projected profits during the tourist season months.

opposed to the categorical data of bar charts. The intervals for a histogram must be created, whereas they are given for a bar chart.

A *pie chart* is a visualization tool for categorical data for which it is important to compare the parts of a whole. This idea is illustrated in Figure 12.12 which depicts the breakdown of the population by ethnic group.

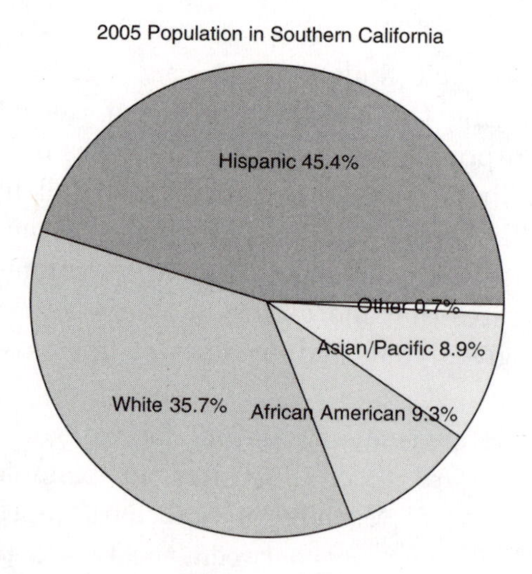

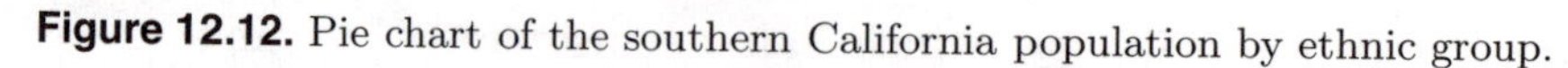

Figure 12.12. Pie chart of the southern California population by ethnic group.

Histograms, bar charts, and pie charts are visualization tools that are found in nearly every field of study; however, there are many other tools. A software package in a particular field of work is the best place to look for examples.

Each of the visualization methods in this section can be extended to 3D.

12.4 Box Plots

Let's suppose a particular scientist runs a number of experiments, each time measuring the same quantity (with slightly changed conditions). A popular way for visualizing such data is the use of *box plots*, also known as *box-and-whiskers-plots*, which summarize the data from each experiment with five numbers. To produce the box

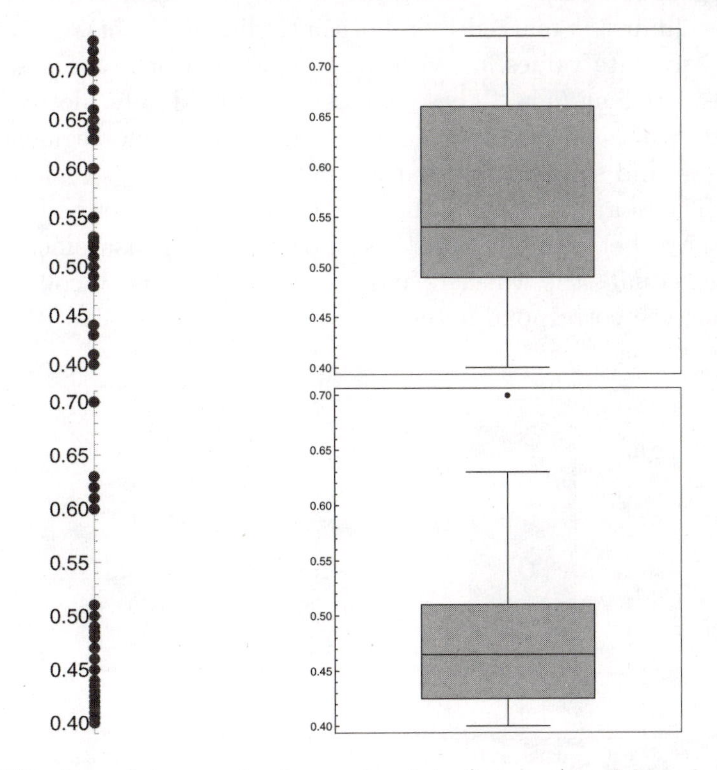

Figure 12.13. Box plot examples for regular data (top row) and biased data (bottom row).

plot for the data, $m_1, \ldots, m_K$, corresponding to one experiment, we use the following steps:

1. Compute the median m of all m_j. This point divides the data into a lower half and an upper half of the m_j.
2. Without using the median, compute the median q_l of the lower half of the m_j. The value q_l is referred to as the *lower quartile.*
3. Again without using the median, compute the median q_u of the upper half of the m_j. The value q_u is referred to as the *upper quartile.*
4. Compute $\Delta = q_u - q_l$, the *interquartile range.*

The box part of the box plot is now a vertical box with horizontal edges at q_u and q_l. Typically, the median m (guaranteed to be within these limits) is marked by a horizontal line segment.

Any data values m_j with $m_j > q_u + 1.5\Delta$ or $m_j < q_l - 1.5\Delta$ are considered *outliers.* These values are individually plotted as dots. The "whiskers" are now added as horizontal line segments at the largest and smallest m_j that are not outliers.

The more variation in the m_j, the taller the boxes and the greater distance between the whiskers. For perfect measurements (all m_j being equal), the whiskers and the three medians all collapse to one value; the corresponding box has no vertical extension.

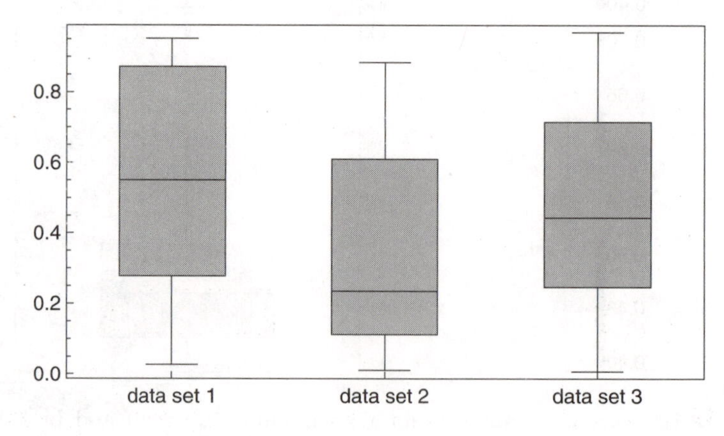

Figure 12.14. Box plots for data sets from three experiments.

Figure 12.13 illustrates two data samples and their box plots. In the top row of the figure the data are rather regular and dense within its interval, and therefore the whiskers are nearly equal in length and there are no outliers identified. In the bottom row, the data are biased to the lower end of the interval, which is reflected by the long whisker extending past the upper quartile. One outlier is identified.

As illustrated in Figure 12.14, box plots for several experiments displayed together concisely describe variations in the data sets.

12.5 Log Plots

Let's investigate a plotting technique that will be very helpful for visualizing empirical data that cover a very wide range of values. Normally we visualize data over the vertical and horizontal axes with a linear spacing. This is called *quadrille ruling.* In some situations, a better plotting method is a *log plot* where one or both axes are given a logarithmic scale.

Suppose we are conducting an experiment that requires noise to be measured at various locations. We measure sound pressure (in Pascals, or Pa), and record the following readings:

	Source of sound	Sound pressure (Pa)
1	Auditory threshold	0.00002
2	Calm breathing	0.00006
3	Very calm room	0.0002
4	Normal talking	0.002
5	Passenger car, 10 m distance	0.02
6	Jackhammer, 1 m distance	2
7	Jet engine, 100 m distance	6

Using a quadrille ruling, a plot of the data appears as illustrated in Figure 12.15 (left). This visualization does not communicate the difference between the first four sounds. Instead, if we plot the pressures as in Figure 12.15 (right), with respect to a base 10 logarithm ($\log_{10}$) scale,[2] we can see the difference between the data points

[2]Recall that $\log_{10}(x) = 10^x$.

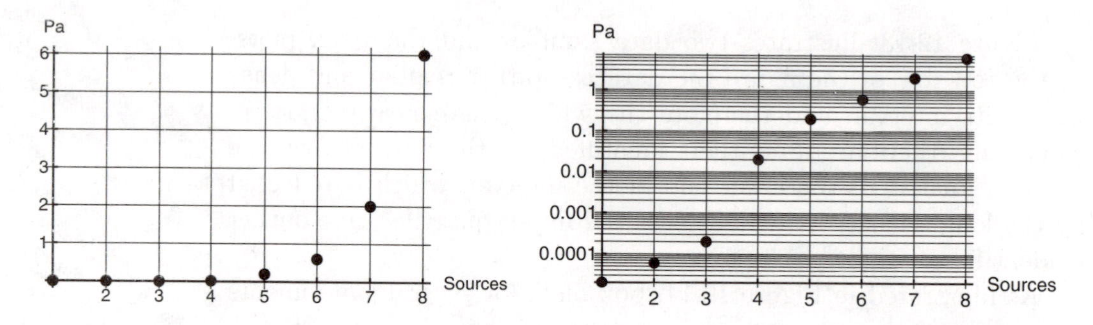

Figure 12.15. Visualization of sampled sound pressure levels. Sound pressures plotted on a quadrille ruling (left) and sound pressures in a log-lin plot (right).

more clearly. This figure is called a *log-lin plot* because the y-axis is marked with a logarithmic scale and the x-axis is marked with a linear scale. Notice that the spacing of the y-axis markings increases as a power of 10 is approached. Each power of 10 that spans the given range of data is given the same partition in the graph. The log-lin plot and similarly the lin-log plot are called *semi-log plots.* A *log-log plot* will scale both axes logarithmically.

Figure 12.15 (right), the log-lin plot, is similar to Figure 12.16 in which the sound pressure values have been converted to decibels, which is a (unitless) $\log_{10}$ scale,

$$x_{dB} = 10 \log_{10}(x_{Pa}/t_{Pa}),$$

where x_{dB} is the decibel value, x_{Pa} is the Pascal pressure value, and t_{Pa} is a threshold (minimum) measurable sound pressure. In fact,

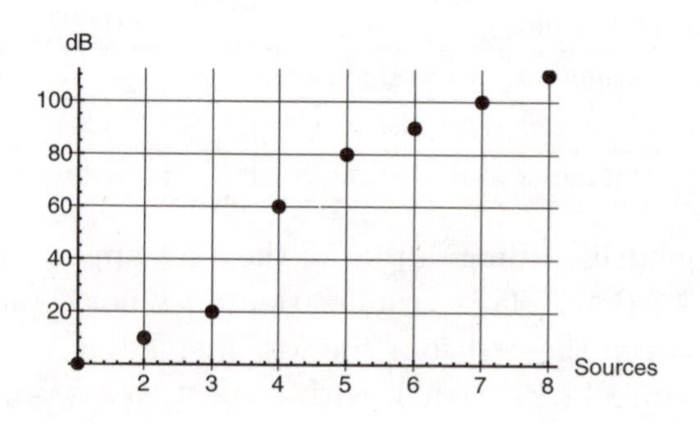

Figure 12.16. Visualization of sound pressure levels sampled which have been converted to decibels.

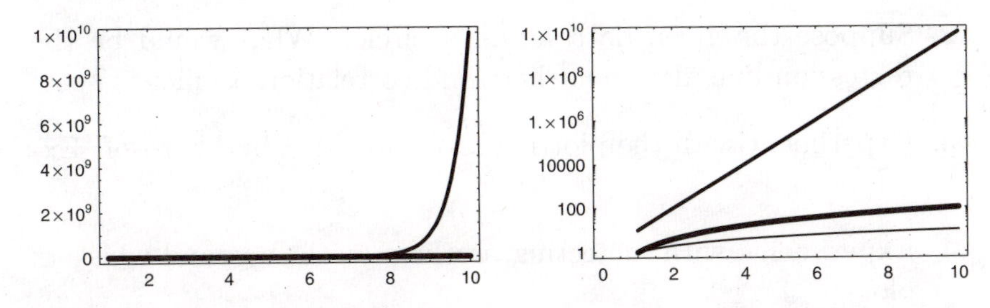

Figure 12.17. Plots of the functions $y = x$, $y = 10^x$, and $y = x^2$ with increasing line thickness, respectively: In the quadrille plot, the exponential function dominates (left), and in the log-lin plot, all three functions are visible (right).

decibles are used for sound partly because the log scale imitates the human perception of loudness. Other fields, in which measured data can span a very large range, also employ logarithmic scales; examples include the Richter magnitude scale for measuring earthquake intensity and the pH (acidity) of a chemical solution.

Figure 12.17 illustrates the effect of the log-lin plot on several representative curves, namely $y = x$, $y = 10^x$, and $y = x^2$. In the log-lin plot (right), the exponential function appears linear and the linear function appears logarithmic. Thus we see that logarithms are very useful for analyzing exponential phenomena such as population growth or decay of isotopes.

Other logarithmic functions, such as the natural log, can be used for log plots as well. Historically, $\log_{10}$ has been use in mathematics and engineering because conveniently the mantissa remains constant and the characteristic increments by one with each power of ten. For example,

$$\log(12) = 1.079, \qquad \log(120) = 2.079, \qquad \log(1200) = 3.079,$$

with truncation.

Log plots should not be used to accommodate outliers in a data set. The box plots from Section 12.4, for example, provided a method for detecting outliers.

12.6 Problems and Experiments

1. Experiment with the relationship between correlation coefficient and slope of the regression line.

2. Suppose the given data lie on a circle. What would be the regression line, dominant line, and correlation coefficient?

3. Experiment with the plotting tools provided by Microsoft Excel.

4. Experiment with detecting outliers with box-and-whiskers plots.

5. Find an application and a data set that needs log-log plots.

13 Facets

Increasingly, 3D data acquisition techniques, such as laser scanners, are used to bring the physical world into a digital form. Powerful computers that are equipped with large amounts of memory can be credited for this increase in use. Laser scanners produce *point clouds*—unorganized 3D points. Analysis and rendering of these point clouds are most efficiently done when the points are connected to form *facets*—planar entities.

13.1 Triangles

The subject of triangles as taught in high school constructed them from angles, edge lengths, and trigonometric functions. Typically, we dealt with *one* triangle. In computing and visualization, we also encounter triangles; however now they exist in rather large numbers. Some objects consist of up to 1,000,000,000 triangles.

We present a brief review of triangle facts. Everything in this review holds for both 2D and 3D triangles.

Triangle. A *triangle* is a 2D or 3D object defined by three vertices $\mathbf{v}_1, \mathbf{v}_2, \mathbf{v}_3$, where the $\mathbf{v}_i$ are points in 2D or 3D.

Triangle edges. The edge opposing vertex i is denoted by E_i; its length is denoted by e_i.

Centroid. The *centroid* $\mathbf{c}$ (center of mass) of a triangle is given by

$$\mathbf{c} = \frac{1}{3}(\mathbf{v}_1 + \mathbf{v}_2 + \mathbf{v}_3). \tag{13.1}$$

Incenter. The *incircle* is the unique circle touching all three triangle edges. Its center, the *incenter* $\mathbf{i}$, is given by

$$\mathbf{i} = \frac{e_1\mathbf{v}_1 + e_2\mathbf{v}_2 + e_3\mathbf{v}_3}{e_1 + e_2 + e_3}.$$

Circumcenter. There is a unique circle, the *circumcircle*, passing through all three $\mathbf{v}_i$. Its *circumcenter* $\mathbf{cc}$ is given by

$$\begin{aligned}\mathbf{cc} = &\frac{(\alpha_1(\alpha_2 + \alpha_3 - \alpha_1))}{S}\mathbf{v}_1 \\ &+ \frac{(\alpha_2(\alpha_3 + \alpha_1 - \alpha_2))}{S}\mathbf{v}_2 \\ &+ \frac{(\alpha_3(\alpha_1 + \alpha_2 - \alpha_3))}{S}\mathbf{v}_3,\end{aligned}$$

where $\alpha_1 = e_1^2$, $\alpha_2 = e_2^2$, $\alpha_1 = e_3^2$, and S is the sum of the coefficient numerators, namely,

$$S = 2\alpha_1\alpha_2 + 2\alpha_1\alpha_3 + 2\alpha_2\alpha_3 - \alpha_1^2 - \alpha_2^2 - \alpha_3^2.$$

The circumcenter is at the intersection of the perpendicular bisectors of the triangle edges.

Normal. The vector $\mathbf{n}$ perpendicular to a triangle, given by

$$\mathbf{n} = \frac{(\mathbf{v}_2 - \mathbf{v}_1) \wedge (\mathbf{v}_3 - \mathbf{v}_1)}{\|(\mathbf{v}_2 - \mathbf{v}_1) \wedge (\mathbf{v}_3 - \mathbf{v}_1)\|}, \tag{13.2}$$

is referred to as the *normal.* If the vertices are 2D, then a $z = 0$ coordinate should be added to the vertices so a cross product is possible. The normal is *normalized,* meaning it is scaled to be of unit length. If two vertices are exchanged in the definition of a triangle—say, for example, the triangle is defined as $\mathbf{v}_1, \mathbf{v}_3, \mathbf{v}_2$—then the normal will flip. Thus, the order implies an orientation for the triangle due to the normal's definition by a cross product, which follows the *right-hand rule.*[1]

---

[1]The right-hand rule is described in Section 3.1.

Area. The *area* $\mathcal{A}$ of a triangle can be given by

$$\mathcal{A} = \sqrt{(\mathbf{v}_2 - \mathbf{v}_1)^2(\mathbf{v}_3 - \mathbf{v}_1)^2 - [(\mathbf{v}_2 - \mathbf{v}_1)(\mathbf{v}_3 - \mathbf{v}_1)]^2}. \quad (13.3)$$

This expression has the nice feature of working for 2D or 3D without modification of the vertices. Another method for calculating the area is

$$\mathcal{A} = 1/2\|(\mathbf{v}_2 - \mathbf{v}_1) \wedge (\mathbf{v}_3 - \mathbf{v}_1)\| \quad (13.4)$$

This method is convenient to use if the normal to the triangle is needed as well. If the vertices are 2D, then a $z = 0$ coordinate should be added to each vertex.

Shape. A triangle is considered to have good *shape* if it is close to being equilateral. Shape may be measured in many ways; one is the distance of the circumcenter to the incenter, which is zero for equilateral triangles.

13.2 Barycentric Coordinates

Let's assume we have a triangle $\mathcal{T}$ with (2D or 3D) vertices $\mathbf{v}_1, \mathbf{v}_2, \mathbf{v}_3$ as well as another point $\mathbf{p}$ in the plane formed by the triangle. A common task is to write $\mathbf{p}$ as a linear combination of the $\mathbf{v}_i$:

$$\mathbf{p} = u_1\mathbf{v}_1 + u_2\mathbf{v}_2 + u_3\mathbf{v}_3, \quad (13.5)$$

meaning that we have to produce the three scalars u_1, u_2, u_3. They are given by

$$u_1 = \text{area}[\mathbf{p}, \mathbf{v}_2, \mathbf{v}_3]/\mathcal{A}, \quad (13.6)$$
$$u_2 = \text{area}[\mathbf{p}, \mathbf{v}_3, \mathbf{v}_1]/\mathcal{A}, \quad (13.7)$$
$$u_3 = \text{area}[\mathbf{p}, \mathbf{v}_1, \mathbf{v}_2]/\mathcal{A}, \quad (13.8)$$

where $\mathcal{A}$ is the area of the triangle. Figure 13.1 provides an illustration.

The area calculations must produce a signed *area*. Assume that the area of $\mathcal{T}$ is positive. The sign of each of the areas in equations (13.6) to (13.8) is determined by comparing the normal vector of

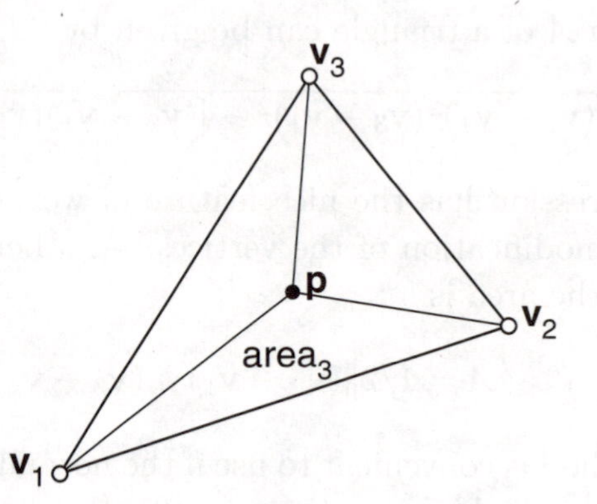

Figure 13.1. Barycentric coordinates form a coordinate frame with respect to a triangle. The label area_3 corresponds to the area for u_3 in (13.8).

these oriented triangles against $\mathcal{T}$'s normal vector of the given triangle. If the normal vectors share the same direction, then the area is positive; otherwise, the area is negative.

In the 2D case, we naturally have

$$u_1 + u_2 + u_3 = 1.$$

In the 3D case, this only holds if, in fact, $\mathbf{p}$ is in the plane formed by the three $\mathbf{v}_i$.

Notice that if $\mathbf{p} = \mathbf{v}_1$, then the barycentric coordinates $[u_1, u_2, u_3] = [1, 0, 0]$. Similarly, if $\mathbf{p} = \mathbf{v}_2$, then the barycentric coordinates are $[0, 1, 0]$, and if $\mathbf{p} = \mathbf{v}_3$, then the barycentric coordinates are $[0, 0, 1]$. The centroid of the triangle has barycentric coordinates $[1/3, 1/3, 1/3]$; this fact was given in (13.1). In fact, the barycentric coordinates for the incenter and circumcenter were given as well. If $\mathbf{p}$ is *inside* the triangle formed by $\mathbf{v}_1, \mathbf{v}_2, \mathbf{v}_3$, then we have $u_i \geq 0$ for all three u_i. Otherwise, we conclude that $\mathbf{p}$ is outside of the triangle. This is typically used as a quick inside/outside test. All points on the edge formed by $\mathbf{v}_2$ and $\mathbf{v}_3$ have barycentric coordinates $[0, u_2, u_3]$, which correspond to the parameters in linear interpolation.

13.3 Planes

The facets of an object are parts of 3D planes. Let's look at two representations of a plane: the implicit and parametric forms. We next examine the best use of each form.

13.3.1 Implicit Equation of a Plane

Suppose we are given a point $\mathbf{p}$ and a vector $\mathbf{n}$ bound to $\mathbf{p}$. The locus of all points $\mathbf{x}$ that satisfy the equation

$$\mathbf{n} \cdot (\mathbf{x} - \mathbf{p}) = 0 \tag{13.9}$$

defines the *implicit form* of a plane. The vector $\mathbf{n}$ is called the *normal* to the plane if $\|\mathbf{n}\| = 1$. If this is the case, then (13.9) is called the *point normal plane equation.*

Expanding (13.9), we have

$$Ax_1 + Bx_2 + Cx_3 + D = 0, \tag{13.10}$$

where

$$\begin{aligned} A &= n_1 \\ B &= n_2 \\ C &= n_3 \\ D &= -(n_1p_1 + n_2p_2 + n_3p_3). \end{aligned}$$

Let's try an example. Compute the implicit form of the plane through the point $\mathbf{p}$ with normal $\mathbf{n}$, where

$$\mathbf{p} = \begin{bmatrix} 4 \\ 0 \\ 0 \end{bmatrix} \qquad \mathbf{n} = \begin{bmatrix} 1 \\ 1 \\ 1 \end{bmatrix}.$$

All we need to compute is $D = -4$. Thus, the plane equation is

$$x_1 + x_2 + x_3 - 4 = 0.$$

Similar to a 2D implicit line, if the coefficients A, B, C correspond to the (unit length) normal to the plane, then $|D|$ describes the distance of the plane to the origin. This is the *perpendicular distance.*

In addition, the point normal form reflects the (perpendicular) distance of a point from a plane. The distance d of an arbitrary point $\hat{\mathbf{x}}$ from the point normal form of the plane is

$$d = A\hat{x}_1 + B\hat{x}_2 + C\hat{x}_3 + D.$$

Suppose we would like to find the distance of many points to a given plane. Then it is computationally more efficient to have the plane in the form of (13.10), corresponding to the point normal form. If we need to know only on which side of the plane a point lies, then it is not necessary to have the implicit form in point normal form; the sign of the result is not changed by the normalization. But keep in mind that the sign of the result is dependent upon the direction of the vector defined by the coefficients A, B, C.

13.3.2 Parametric Equation of a Plane

The implicit plane equation is wonderful for determining whether a point is in a plane; however, it is not so useful for creating points in a plane. For this, we have the *parametric form* of a plane.

The *given* information for defining a parametric representation of a plane usually comes in one of two ways:

- three points, or
- a point and two vectors.

If we start with the first scenario, we choose three points $\mathbf{p}, \mathbf{q}, \mathbf{r}$, then choose one of these points and form two vectors $\mathbf{v}$ and $\mathbf{w}$ bound to that point:

$$\mathbf{v} = \mathbf{q} - \mathbf{p} \quad \text{and} \quad \mathbf{w} = \mathbf{r} - \mathbf{p}. \tag{13.11}$$

Why not just specify one point and a vector in the plane, analogous to the implicit form of a plane? This is not enough information to uniquely define a plane. Many planes fit that data.

Two vectors bound to a point are the data we'll use to define a plane $\mathbf{P}$ in parametric form as

$$\mathbf{P}(s, t) = \mathbf{p} + s\mathbf{v} + t\mathbf{w}. \tag{13.12}$$

The two independent parameters, s and t, determine a point $\mathbf{P}(s,t)$ in the plane.[2] Notice that (13.12) can be rewritten as

$$\begin{aligned}\mathbf{P}(s,t) &= \mathbf{p} + s(\mathbf{q} - \mathbf{p}) + t(\mathbf{r} - \mathbf{p}) \\ &= (1 - s - t)\mathbf{p} + s\mathbf{q} + t\mathbf{r}.\end{aligned} \tag{13.13}$$

The coordinates $(1 - s - t, s, t)$ are the *barycentric coordinates* of a point $\mathbf{P}(s,t)$ with respect to the triangle with vertices $\mathbf{p}, \mathbf{q}$, and $\mathbf{r}$. See Section 13.2 for a complete discussion of barycentric coordinates.

Another method for specifying a plane is as the *bisector of two points.* This is how a plane is defined in Euclidean geometry—the locus of points equidistant from two points. The line between two given points defines the normal to the plane, and the midpoint of this line segment defines a point in the plane. With this information it is most natural to express the plane in implicit form.

13.3.3 Intersecting Three Planes

Suppose we are given three planes with implicit equations

$$\begin{aligned}\mathbf{n}_1 \cdot \mathbf{x} + c_1 &= 0, \\ \mathbf{n}_2 \cdot \mathbf{x} + c_2 &= 0, \\ \mathbf{n}_3 \cdot \mathbf{x} + c_3 &= 0.\end{aligned}$$

Where do they intersect? The answer is at some point $\mathbf{x}$, which lies on each of the planes.

The solution is surprisingly simple; just condense the three plane equations into matrix form:

$$\begin{bmatrix}\mathbf{n}_1^{\mathrm{T}} \\ \mathbf{n}_2^{\mathrm{T}} \\ \mathbf{n}_3^{\mathrm{T}}\end{bmatrix} \begin{bmatrix}x_1 \\ x_2 \\ x_3\end{bmatrix} = \begin{bmatrix}-c_1 \\ -c_2 \\ -c_3\end{bmatrix}. \tag{13.14}$$

We have three equations in the three unknowns x_1, x_2, x_3! Chapter 5 identifies methods to solve such linear systems.

Let's find the intersection of the following three planes,

$$x_1 + x_3 = 1, \quad x_3 = 1, \quad x_2 = 2.$$

[2]This is a slight deviation in notation: an uppercase boldface letter rather than a lowercase one denoting a point.

The linear system is

$$\begin{bmatrix} 1 & 0 & 1 \\ 0 & 0 & 1 \\ 0 & 1 & 0 \end{bmatrix} \begin{bmatrix} x_1 \\ x_2 \\ x_3 \end{bmatrix} = \begin{bmatrix} 1 \\ 1 \\ 2 \end{bmatrix}.$$

Solving by Gauss elimination, we obtain

$$\begin{bmatrix} x_1 \\ x_2 \\ x_3 \end{bmatrix} = \begin{bmatrix} 0 \\ 2 \\ 1 \end{bmatrix}.$$

While simple to solve, the three-planes problem does not always have a solution. If the normal vectors $\mathbf{n}_1, \mathbf{n}_2, \mathbf{n}_3$ are linearly dependent, then there is no solution to the intersection problem. An example will illustrate.

The normal vectors are

$$\mathbf{n}_1 = \begin{bmatrix} 1 \\ 0 \\ 0 \end{bmatrix}, \quad \mathbf{n}_2 = \begin{bmatrix} 1 \\ 0 \\ 1 \end{bmatrix}, \quad \mathbf{n}_3 = \begin{bmatrix} 0 \\ 0 \\ 1 \end{bmatrix}.$$

Since $\mathbf{n}_2 = \mathbf{n}_1 + \mathbf{n}_3$, they are indeed linearly dependent, and thus the planes defined by them do not intersect in one point.

13.4 Polygons and Polyhedra

A triangle is a planar object (in 2D or 3D) having three vertices. If we allow for more vertices, then we have a *polygon.* A collection of more than three 3D points need not be planar, but for them to form a polygon, we do require coplanarity, meaning all points lie in one plane, having a normal $\mathbf{n}$. Polygons may self-intersect. If they do not, they are called *simple.*

Among the simple polygons, *convex* ones play an important role (see e.g. Voronoi diagrams in Section 13.8). A polygon is convex if no straight line intersects it more than twice. Figure 13.2 shows some examples.

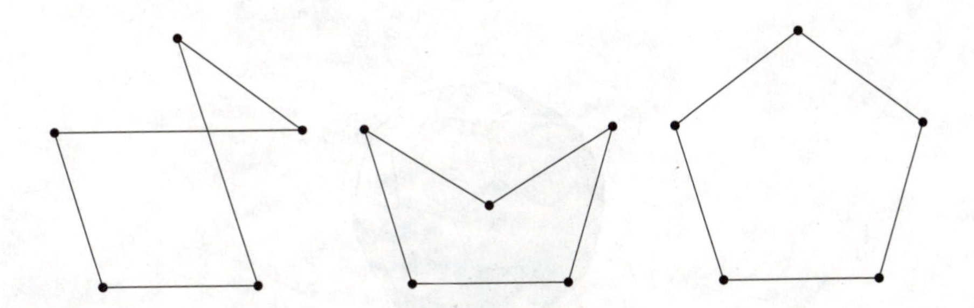

Figure 13.2. Examples of polygons: non-simple (left), simple (middle), and convex (right).

The *area* A of a 2D simple polygon with p vertices $\mathbf{x}_1, \ldots, \mathbf{x}_p$ is given by

$$A = \frac{1}{2}[x_1 y_2 - y_1 x_2 + \ldots + x_p y_1 - y_p x_1].$$

The area of a 3D simple polygon with 3D vertices $\mathbf{x}_1, \ldots, \mathbf{x}_p$ and normal vector $\mathbf{n}$ is given by

$$A = \frac{1}{2}\mathbf{n}[\mathbf{x}_1 \wedge \mathbf{x}_2 + \ldots + \mathbf{x}_p \wedge \mathbf{x}_1]$$

where $\wedge$ is the 3D cross product. Note that we form a dot product with $\mathbf{n}$ such that the result is a real number! If we flip the normal and use $-\mathbf{n}$ instead, then the area changes sign.

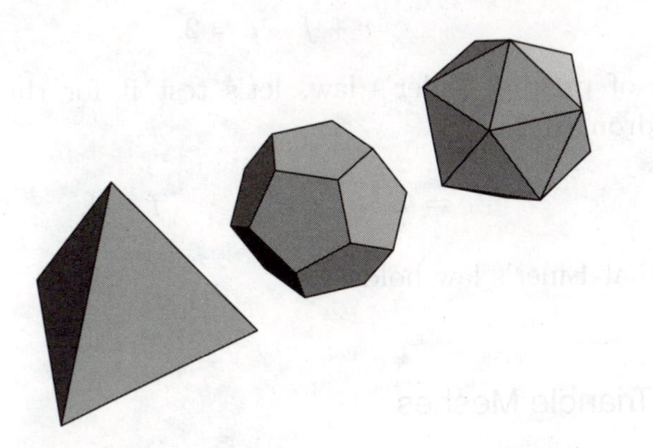

Figure 13.3. Examples of polyhedra: tetrahedron (left), dodecahedron (middle), and icosahedron (right).

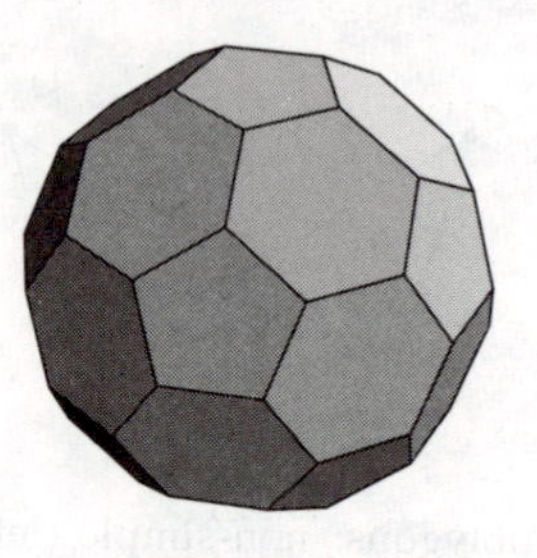

Figure 13.4. A buckyball.

A 3D object whose faces are convex polygons is known as a *polyhedron*. Classic examples are tetrahedra and cubes. Figure 13.3 shows some examples. Polyhedra play an important role in *crystallography*: a crystal has the internal atomic packing structure of polyhedra and often exhibits polygonal facets in larger physical specimens.

Another famous polyhedron, illustrated in Figure 13.4, derives from chemistry: the *buckyball*. The buckyball is the complex carbon molecule C_{60}, consisting of 60 carbon atoms arranged in the shape of a soccer ball. The name derives from Buckminster Fuller, an architect whose claim to fame was the geodesic dome reminiscent of the buckyball shape.

All polyhedra obey *Euler's law*: if there are v vertices, e edges, and f faces, then

$$v + f - e = 2.$$

Instead of proving Euler's law, let's test it for the example of a tetrahedron. Here,

$$v = 4, \qquad e = 6, \qquad f = 4.$$

Check that Euler's law holds!

13.5 Triangle Meshes

A *triangle mesh* or *triangulation* is an object formed by triangular facets; these could all be in the xy-plane or they could be in 3D.

An example of a 2D triangle mesh is a square with one diagonal added; a 3D example is given by the four triangles on the outside of a tetrahedron.

Triangle meshes became popular with the widespread use of 3D data acquisition techniques, such as laser scanners, stereo photogrammetry, touch probes, 3D satellite images, confocal scanning laser microscopes, and more. All of these techniques extract a collection of 3D points from an object (from a mountain range for satellites to a blood cell for microscopes). These point collections are then organized into triangle meshes.

The other major source of triangle meshes is from finite element computations. Here, an object (2D or 3D) is broken down into triangles and a PDE is solved discretely, similar to the method presented in Section 11.6.

In general, a triangle mesh is given by a collection of triangles T_j such that any two triangles either have no points in common or they share exactly one common edge or exactly one common vertex. Figure 13.5 shows a valid and invalid 2D triangle mesh.

Triangle meshes are collections of triangles, each consisting of three vertices and three edges. This trivial fact leads the way to defining a *data structure*, which is used by most standard formats. A triangle mesh is defined by its *geometry* (i.e., by the set of its vertices) and by its *connectivity* (i.e., a list of triangles). For a simple

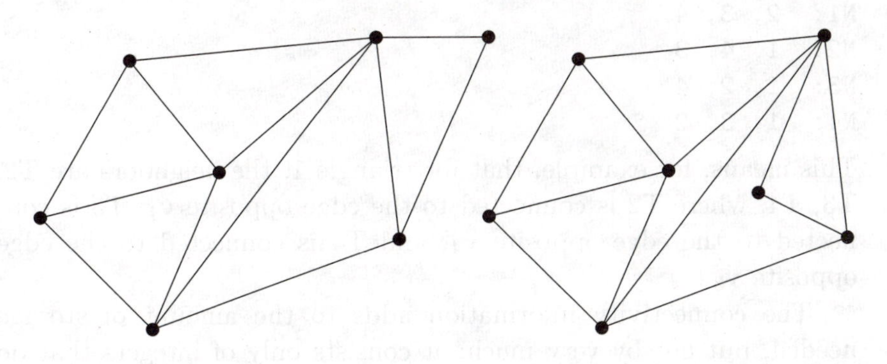

Figure 13.5. A valid 2D triangle mesh (left), in contrast to an invalid one (right).

example, let's take a tetrahedron-like mesh with vertices

$$\mathbf{v}_1 = \begin{bmatrix} 0 \\ 0 \\ 0 \end{bmatrix}, \quad \mathbf{v}_2 = \begin{bmatrix} 0 \\ 0 \\ 1 \end{bmatrix}, \quad \mathbf{v}_3 = \begin{bmatrix} 1 \\ 0 \\ 0 \end{bmatrix}, \quad \mathbf{v}_4 = \begin{bmatrix} 0 \\ 1 \\ 0 \end{bmatrix}.$$

The corresponding data structure would start with a floating-point vertex list as follows.

```
v1:  0.0, 0.0, 0.0
v2:  0.0, 0.0, 1.0
v3:  1.0, 0.0, 0.0
v4:  0.0, 1.0, 0.0
```

Following this list will be information about which vertices form a triangle. For this example, we have the following integer triangle list, pointing to the vertex list.

```
T1:  1,4,3
T2:  2,3,4
T3:  2,1,3
T4:  2,4,1
```

This means, for example, that triangle T1 is formed by vertices $\mathbf{v}_1, \mathbf{v}_4, \mathbf{v}_3$. All triangles are defined consistently so that outward normals are produced. A third list would give information about the *neighbors* of a triangle. In our example, we would have the following integer neighbor list, pointing to the integer triangle list.

```
N1:  2, 3, 4
N2:  1, 4, 3
N3:  1, 2, 4
N4:  1, 3, 2
```

This means, for example, that for triangle 1, the neighbors are T2, T3, T4, where T2 is connected to the edge opposite $\mathbf{v}_1$, T3 is connected to the edge opposite $\mathbf{v}_4$, and T4 is connected to the edge opposite $\mathbf{v}_3$.

The connectivity information adds to the amount of storage needed, but not by very much: it consists only of integers that do not need much space. The benefit of this additional information can be huge when computing with meshes. A typical operation involves finding all neighbors of a given triangle. This is trivial when using the data structure above, but would amount to searching the whole mesh without it!

Most edges in a mesh are shared by two triangles, but some are not. A triangle edge that is only part of one triangle—it is not shared by two triangles—is called a *boundary edge.* In a data structure, if the edge across from a vertex is a boundary edge, then a -1 (or other nonvalid triangle index) is stored as the pointer to a (nonexistent) neighbor triangle. A mesh without boundary edges is called *closed.*

Closed meshes could have the shape of a sphere, a tooth, or a pebble. Examples of more complex shapes are a torus (or doughnut) or a human skull. If covered by 3D meshes, there will not be any boundary edges, and yet a torus as well as a skull have *holes.* The number of holes is called the *genus* of the mesh. A sphere has genus zero; a torus has genus one.

At any vertex $\mathbf{v}_i$ of a triangle mesh, we may define a *normal vector*, which is sometimes called a *vertex normal.* This is a unit vector roughly perpendicular to the mesh at $\mathbf{v}_i$, and is important in applications such as rendering; see Section 16.3. Let each of the $j = 1, N$ triangles with $\mathbf{v}_i$ as a vertex have a normal vector $\mathbf{n}_{i,j}$. Then a reasonable guess for the normal $\mathbf{n}_i$ at $\mathbf{v}_i$ is the average of all $\mathbf{n}_{i,j}$:

$$\mathbf{n}_{i,j} = \frac{[\mathbf{n}_{i,1} + \ldots + \mathbf{n}_{i,N}]}{\|[\mathbf{n}_{i,1} + \ldots + \mathbf{n}_{i,N}]\|}. \tag{13.15}$$

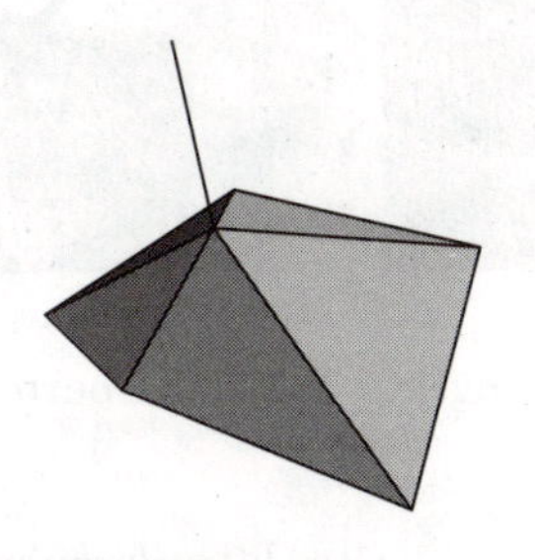

Figure 13.6. A normal for a vertex in a 3D mesh.

Because the vertex normal is normalized to be of unit length, the factor $1/N$ can be ignored. An example is shown in Figure 13.6.

See Sections 14.1 and 15.1 for visualization techniques for meshes and more applications.

13.6 Case Study: 3D Archiving

In 1999, a group of Stanford researchers, led by M. Levoy, set out to digitally archive some of Italy's most prized art treasures: statues of Michelangelo.[3] Their tools were 3D laser digitizers that, by shooting laser rays at an object, collect x, y, z-coordinates of points on the object. Once collected, this information can be used as a digital record of the statue. This record is meant to preserve an object's geometry even if the original should be damaged or destroyed.

The laser digitizer simply reads off points from an object. In order to work with that information, structure has to be given to those points. That structure is the triangle mesh. In the case of the David, over a billion triangles were needed, the results of more than 40 scans of different parts of the statue. One such scan in action is shown in Figure 13.7. Figure 13.8 shows part of David's face on the left and part of the triangle mesh for his right eye on the right.

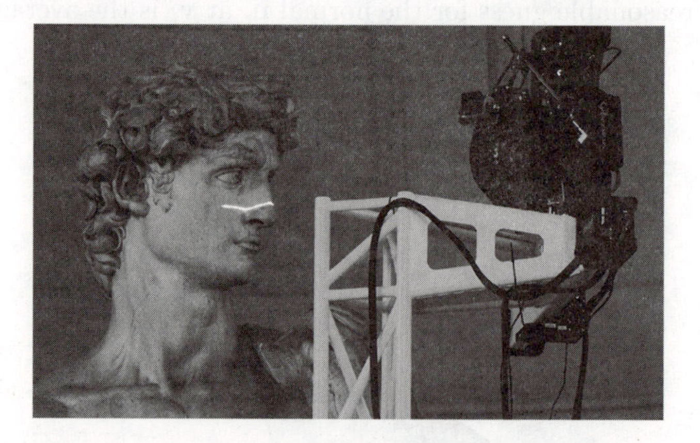

Figure 13.7. Scanning Michelangelo's David. (Courtesy of M. Levoy, Stanford University.)

[3]See the Digital Michelangelo site http://graphics.stanford.edu/projects/mich/ for more details.

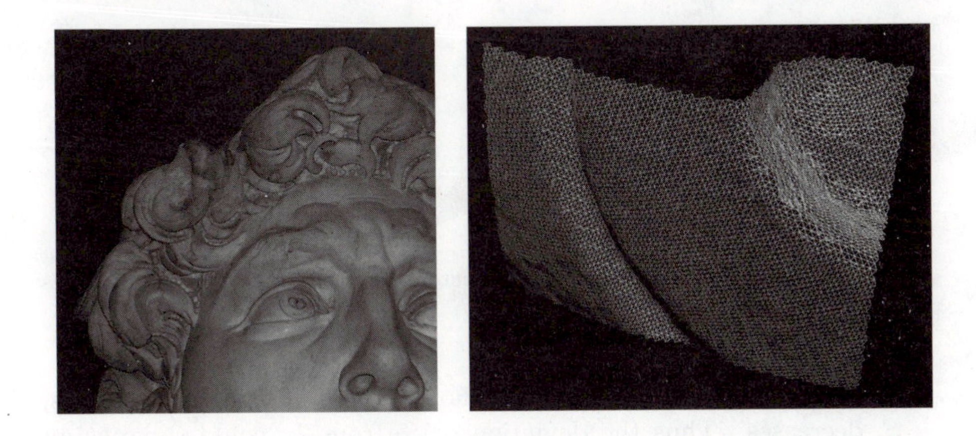

Figure 13.8. Details of Michelangelo's David. (Courtesy of M. Levoy, Stanford University.)

Having the 3D mesh description of an object allows more than just archiving. In the case of the David, art historians had wondered whether Michelangelo had followed the rule that says the statue's center of gravity, when projected downward, needs to be exactly between the two feet. Some simple calculations showed that this is indeed the case here! This fact was impossible to prove before the creation of the digital model.

13.7 Analyzing Triangle Meshes

Suppose a scientist has a 3D mesh of a fossil bone, obtained from scanning the original by using a laser scanner. It would be of interest to extract shape properties and answer questions such as: Are there flat regions? Where do we see highly curved regions?

The key to analyses such as those is our ability to explore local shape measures for a mesh. For that, we define the *star* $\mathbf{v}_i^\star$ of a vertex $\mathbf{v}_i$ to be the set of all triangles having $\mathbf{v}_i$ as a vertex. All three examples in Figure 13.9 show the star of the center vertex. We denote by $|\mathbf{v}_i^\star|$ the area formed by all triangles in $\mathbf{v}_i^\star$.

What can we say about the shape at $\mathbf{v}_i$? In Figure 13.9, the left mesh is completely flat; as we move right, it becomes increasingly pointy. How can we quantify this? Let S_i be the sum of the angles

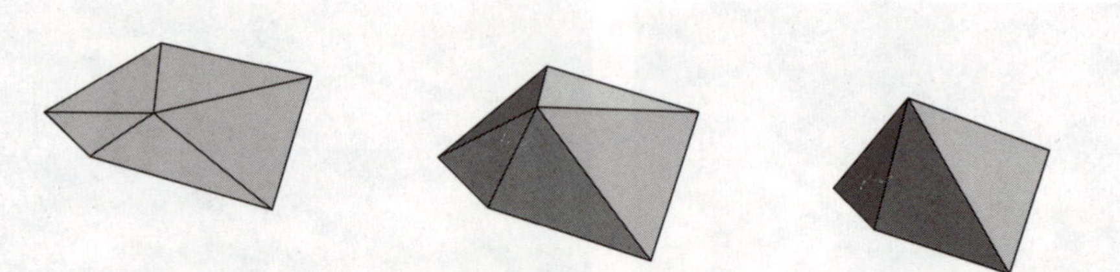

Figure 13.9. The local curvature at the center vertex in the three meshes increases from left to right.

at $\mathbf{v}_i$. In the left case, $S_i = 2\pi$. For the middle and right meshes, S_i decreases. Thus the deviation of S_i from 2π could serve as an indicator of how curved the mesh is at $\mathbf{v}_i$. However, this is not what we intuitively expect from a curvature. When we scale an object up, its curvatures should become smaller (think of circles), but so far we are using only angles, and they do not change under scalings. A solution is given by

$$K_i = \frac{2\pi - S_i}{3|\mathbf{v}_i^\star|}. \tag{13.16}$$

We denote this shape as the *discrete Gauss curvature.* Since we divide by an area, this curvature will decrease as we scale up a mesh.

Curvatures may also be negative, namely, if $2\pi - S_i < 0$. That this can happen is demonstrated in Figure 13.10. Vertices with negative curvatures are called *saddle-shaped*, a notion easily supported by Figure 13.10.

We may use curvatures to *segment* triangle meshes: a segment or region would be characterized by having all of its curvatures in a similar range. One algorithm to achieve this is known as *region*

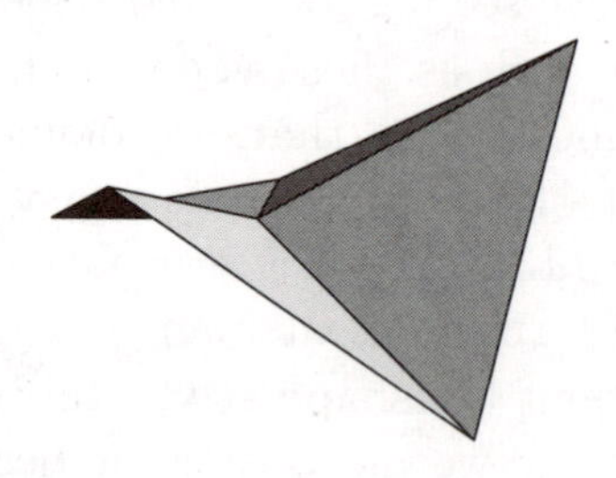

Figure 13.10. Negative curvature at the center vertex.

growing, in which a random vertex (the seed) is selected and its curvature is computed. If its neighbors have similar curvatures, add them to form a region. Continue adding neighbors in this fashion until no more points of similar curvature can be found. Once a region is finished, randomly select another seed and repeat until the whole mesh is covered. The term "similar" involves a threshold tolerance—it will be application dependent. See also the related 2D version presented in Section 14.3.

13.8 Delaunay Meshes and Voronoi Diagrams

Suppose we have a set of 2D points $\mathbf{v}_1, \ldots, \mathbf{v}_L$. Can we make them the vertices of a triangle mesh? The answer is yes—in fact there are many ways to connect points to form triangle meshes. The most popular mesh is known as the *Delaunay triangulation*, or *Delaunay mesh*. It is defined by the *empty circumcircle* property: for any triangle T_i in a Delaunay mesh, no point is inside T_i's circumcircle. In general, only T_i's vertices will be on its circumcircle; however, more than three points can be cocircular. In this case, any triangulation of the cocircular points is allowed.

Why is this good? Let's consider the mesh on the left of Figure 13.5. Four of its points, forming two triangles, are repeated in the left part of Figure 13.11. For each, the circumcircle is shown. (One circumcircle is only partly drawn because it is huge). Each circumcircle contains one other vertex—these two triangles cannot appear

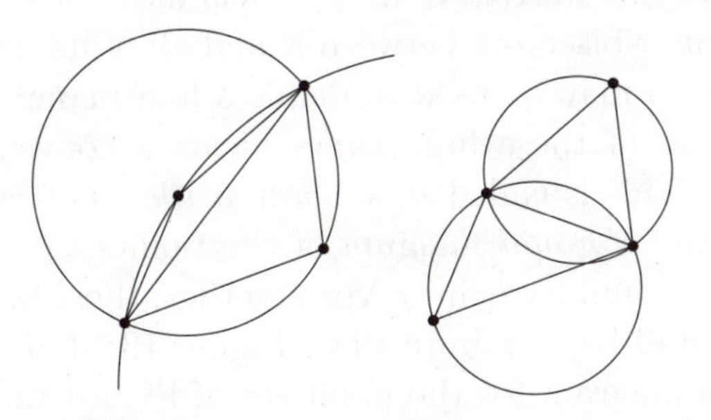

Figure 13.11. Triangle configurations with circumcircles: a non-Delaunay configuration (left), and a Delaunay configuration (right) for the same four points.

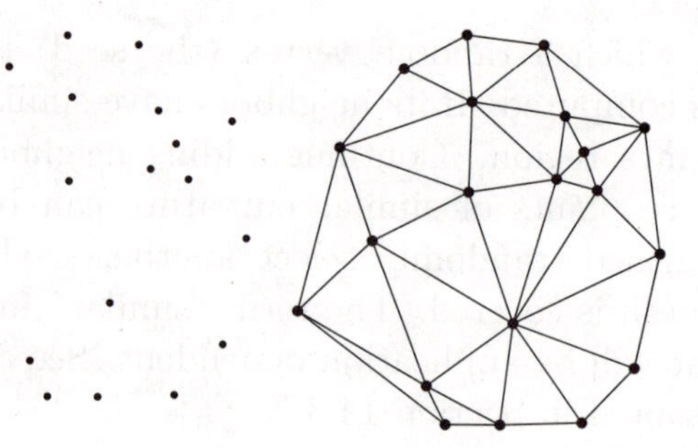

Figure 13.12. A 2D point set (left)and its Delaunay mesh (right).

in a Delaunay mesh. In the right part of Figure 13.11, the same four points are assembled into two different triangles. Their triangle circumcircles do not contain a nontriangle vertex; hence, this is a Delaunay configuration. Intuitively, the shape of the two triangles on the right is superior to that on the left. Thus, the empty circumcircle criterion seems to agree with our sense of "good triangle shape."

Most algorithms construct a Delaunay mesh that covers the *convex hull* of the 2D data set. To understand the concept, think of the points $\mathbf{v}_i$ as nails in a board. We loosely enclose this set of nails by a string, and pull the string tight. What is inside it is the convex hull of the $\mathbf{v}_i$. In practice, badly shaped triangles often appear near the boundary of the convex hull. Figure 13.12 shows a 2D point set (left) and its Delaunay triangulation (right).

Related to Delaunay meshes are *Voronoi diagrams*, constructed as follows. For any vertex $\mathbf{v}_i$ in the Delaunay mesh, we construct the perpendicular bisectors between it and all of its neighbors. Each of these bisectors may be used to define a half-plane[4] containing $\mathbf{v}_i$. The intersection of these half-planes yields a convex polygon (see Section 13.4). This is called $\mathbf{v}_i$'s *Voronoi tile.* The collection of all these tiles is the *Voronoi diagram* of the points $\mathbf{v}_1, \ldots, \mathbf{v}_L$. Points inside the convex hull have finite Voronoi tiles; those on the boundary of the convex hull have infinite tiles. Figure 13.13 gives an example of the Voronoi diagram for the point set of Figure 13.12.

[4]The bisector separates the point set's plane into two parts; a half-plane is one of these parts.

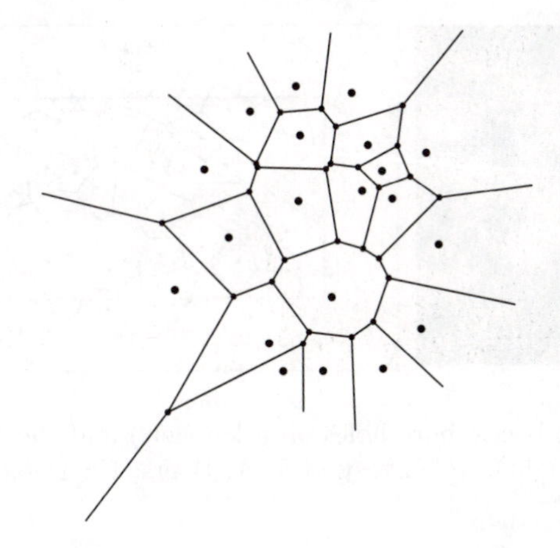

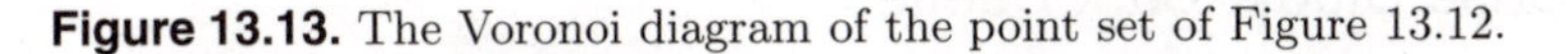

Figure 13.13. The Voronoi diagram of the point set of Figure 13.12.

By the definition of a circumcircle (its center being the intersection of the three edge bisectors), the vertices of the Voronoi tiles are the circumcenters of the triangles in the Delaunay mesh.

The Voronoi tile of a point $\mathbf{v}_i$ consists of parts of the bisectors between $\mathbf{v}_i$ and its neighbors. It follows that all points inside $\mathbf{v}_i$'s tile are closer to $\mathbf{v}_i$ than to any other of the points $\mathbf{v}_k$ for $k = 1, L$ and $k \neq i$.

This fact accounts for the versatility of Voronoi diagrams in many applications. As an early example, A. Thiessen used them in the field of climatology. In a given region, weather stations measure daily rainfall data. Each measurement is exact only for the corresponding weather station. To approximate rainfall at any given location, it makes sense to assign to it the measurement of the closest weather station. Hence for each weather station, its area of validity is its Voronoi tile.[5]

Another application involves cell phone usage. A person traveling in a car and using a cell phone will automatically be switched from one transmitter to another when necessary. Where will this happen? At the boundaries of the transmitters' Voronoi cells! Thus, in the layout of cell phone networks, Voronoi diagrams are an indispensable staple.

[5]Also known by the term Thiessen polygon.

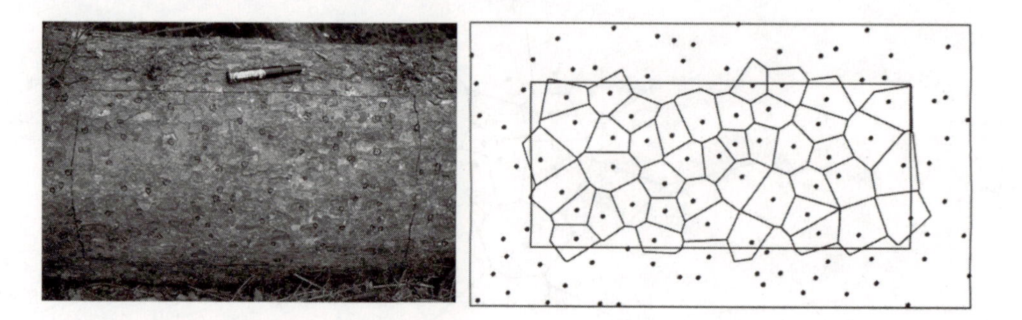

Figure 13.14. Bark beetle bore holes on a log (left) and the resulting bore hole Voronoi diagram (right). (Courtesy of J. A. Byers, US Department of Agriculture.)

13.9 Case Study: Bark Beetles

The forests of the American West as well as those of southern Europe have recently seen a rapid increase in wildfires. This increase is partly due to drought and overall temperature increases. Both of these conditions are amenable to fire conditions, but another important one is the spread of the *bark beetle.* This beetle attacks the barks of pine trees, resulting in the death or weakening of the trees, in turn resulting in unusually high losses due to fires.

Scientists are researching the patterns of bark beetle infestation. It turns out that they do not randomly attack a tree—this would result in a very uneven beetle distribution on a typical bark. Instead, once one beetle has bored a hole into the bark, it releases a pheromone component that lets other beetles know not to attack in the vicinity. As a result, the beetle distribution on a bark is governed by pheromone influence areas, naturally modeled by Voronoi diagrams. Figure 13.14 shows a Voronoi diagram computed from a sample of bark beetle bore holes. A careful analysis (omitted here) shows that the variation in size of the Voronoi tiles is much less than expected from a totally random beetle infestation, showing that the beetles' pheromone distribution tactic is maximizing their damage to the tree.

13.10 Other Meshes

Triangle meshes are not the only types of meshes. For bivariate function representation and contouring of these functions, *rectan-*

Figure 13.15. Rectangle mesh used to realize a glass train station design. (Courtesy of Geometric Modeling and Industrial Geometry, Technical University Vienna.)

gle and *quadrilateral meshes*[6] are widely used. In the domain of the function, say the xy-plane, a rectangular mesh is constructed. Each rectangle is planar. A function value is defined at each vertex in the mesh. The graph of this function over the rectangle mesh forms a quadrilateral mesh since the faces of this mesh are not planar in general. Section 11.4 explores bivariate functions and contouring. Visualization aspects of these topics are continued in Chapter 14.

It is possible to have rectangle meshes in 3D. This would be a restricted mesh, which can occur, for example, in architecture when working with special materials. Figure 13.15 illustrates a train station that is to be built of rectangular glass pieces, which are planar and cannot be deformed. More information on this type of problem may be found in [14].

[6]Quadrilateral meshes often go by the name quad meshes.

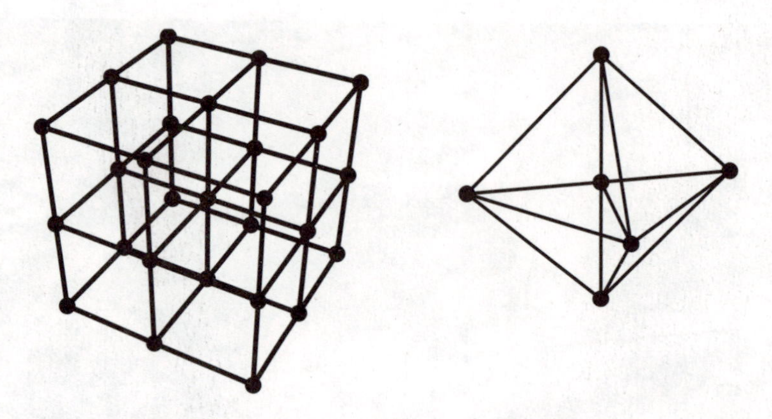

Figure 13.16. 3D meshes: rectilinear grid (left) and tetrahedral mesh (right).

Types of 3D meshes include *rectilinear grids* and *tetrahedral meshes*, illustrated in Figure 13.16. At the vertices of these meshes, function values are often defined. These values might be measured data or a known trivariate function. Chapter 15 presents several motivations for visualizing data over rectilinear or tetrahedral grids are given. A mathematical motivation with respect to trivariate functions is given in Section 11.7.

13.11 Problems and Experiments

1. What are the barycentric coordinates of the incenter of a triangle?
2. Section 13.1 presented one measure for the shape of a triangle. There are many more. Try to find some and compare.
3. Find out how many faces, edges, and vertices a buckyball has. Then confirm Euler's law for the buckyball.
4. The right image in Figure 13.5 is an invalid mesh. Why?
5. Section 13.6 mentioned computing the center of gravity of a triangulated object. An easy solution appears to be to simply take the average of all vertices. Why does that not work? Hint: experiment with points sampled from a sphere, and then put many more points near the north pole.

6. Every set of 2D points defines a Voronoi diagram consisting of a collection of convex polygons. Conversely, does every collection of convex 2D polygons define a set of points of which it is the Voronoi diagram?

14

Visualizing Scalar Values over 2D Data

The following sections look at visualization of scalar values over points forming *a rectilinear grid (gridded)* and *scattered 2D point sets*, as illustrated in Figure 14.1. These data could represent locations over which elevations are known. Or, the data could represent the number of units sold for two toy products over which the profit will be plotted. Perhaps we have a function such as $f(x, y) = \sin(xy)$, that we want to visualize.

In each of these examples, we will be visualizing a scalar function of two variables. This is also called $2\frac{1}{2}D$ *data* because every scalar value is associated with a 2D point. We differentiate this type of data

Figure 14.1. 2D gridded data (left) in contrast to 2D irregular data (right).

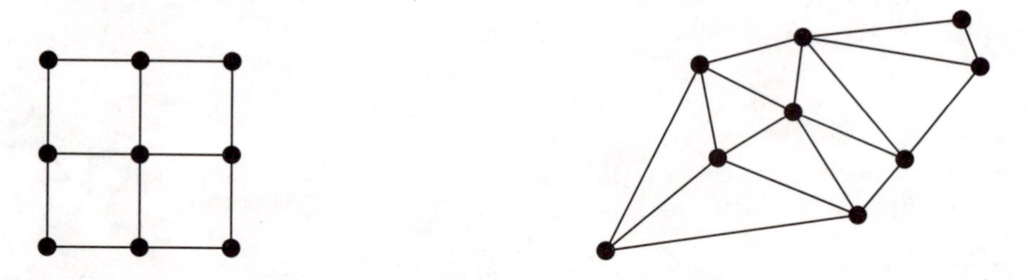

Figure 14.2. Examples of a 2D rectangle mesh (left) and a 2D triangle mesh (right).

from true 3D meshes as discussed in Section 13.5; however, in some fields this distinction is lost. To simplify the following discussion, let's assume the 2D point set lives in the xy-plane. This can always be achieved by an affine map.

A computed tomography (CT) scan returns a picture of a slice through the body. This picture is discretized into pixels, and at each vertex in the pixel array, a radiodensity is reported that will indicate the type of tissue there. This is an example of empirical data on a grid. Another example of gridded data are *digital elevation models* (DEMs), which are output from remote sensing technologies, such as light detection and ranging (LiDAR), for mapping terrains. And yet another example of gridded data occurs when we *evaluate* a bivariate surface (see Section 11.1) over a grid.

A scattered 2D data set might come from weather stations located in a metropolitan area for the purpose of recording rainfall and other weather-related events. Also, tactile digitizing methods will result in scattered data sets.

The visualization techniques we discuss in the sections to follow require a connectivity to be associated with the 2D point set. Chapter 13 discusses some details of meshes. Figure 14.2 illustrates the two types of meshes of interest here: *rectangle and triangle meshes.*

Visualization of vector fields over 2D data is an important topic as well. For bivariate data, this topic is introduced in Section 11.1 in the context of gradient fields. Vector field visualization is introduced in Section 15.9

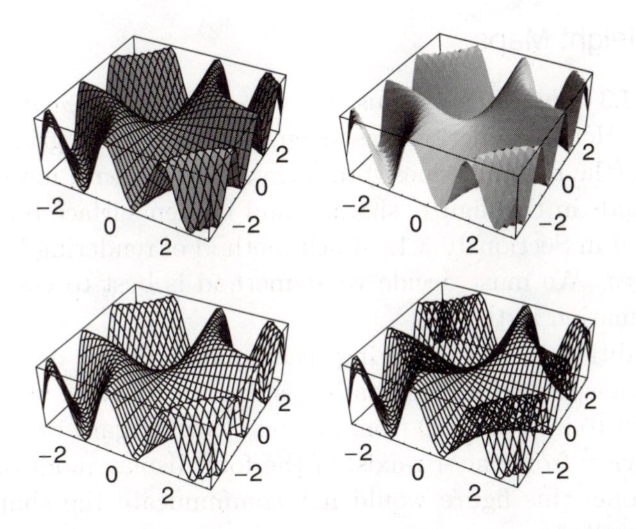

Figure 14.3. Various display modes for visualization of $f(x, y) = \sin(xy)$ over $[-\pi, \pi] \times [-\pi, \pi]$ with 35 evaluations in each axis-direction. Clockwise from the top left: shaded with wireframe, shaded, wireframe, wireframe with hidden surface removal.

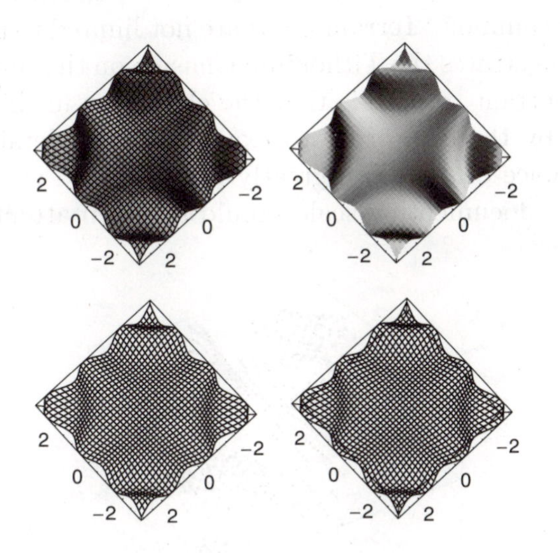

Figure 14.4. Various display modes for visualization of $f(x, y) = \sin(xy)$ over $[-\pi, \pi] \times [-\pi, \pi]$ with 35 evaluations in each axis-direction, viewed from the $+z$-axis, looking into the xy-plane. Each rendering corresponds to the respective display mode in Figure 14.3.

14.1 Height Maps

Figure 14.3 illustrates various ways to display a *height map* with a mesh. Here a function has been evaluated over gridded (x, y) locations (the rectangle mesh) to form a quadrilateral mesh. Two of the methods in this figure, shading and hidden surface removal, are introduced in Section 16.3.1. Each method of rendering has its own advantages. We must decide what method is best to communicate the information in the plot.

In addition to the rendering technique, we must determine the optimal viewpoint if a presentation is limited to static views. It can be difficult to interpret 3D from just one view. Figure 14.4 is a "top" view, or view from the $+z$-axis, of the four display modes in Figure 14.3. Alone, this figure would not communicate the shape of the function well.

Meshes allow us to clearly communicate function values at the 2D point set. Figure 14.5 demonstrates that it is very difficult to make sense of the plot without a mesh structure. The data displayed is called a *point cloud.*

With advances in data acquisition methods, terrain maps have become very common. Terrain maps are not limited to planet Earth. Figure 14.6 illustrates the Tithonium Chasma on the surface of Mars. What is important to note is that the elevations in this figure have been scaled by three in order to exaggerate the detail of the surface. The choice of scale can greatly influence the visualization, so appropriately documenting scale should be given attention. The ef-

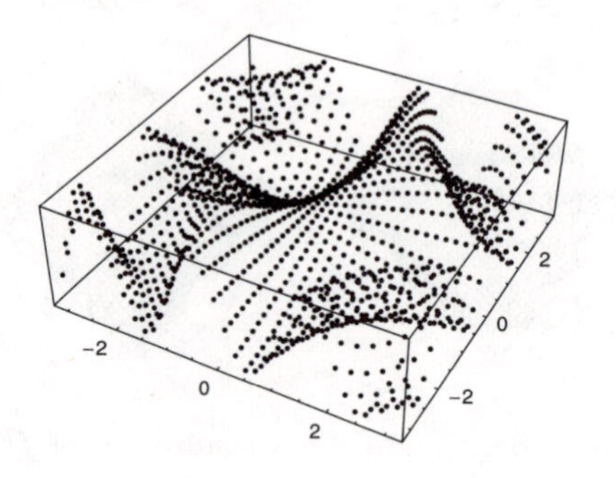

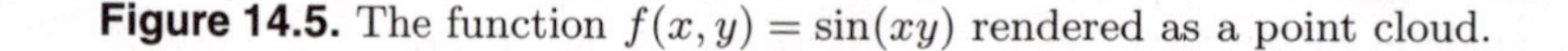

Figure 14.5. The function $f(x, y) = \sin(xy)$ rendered as a point cloud.

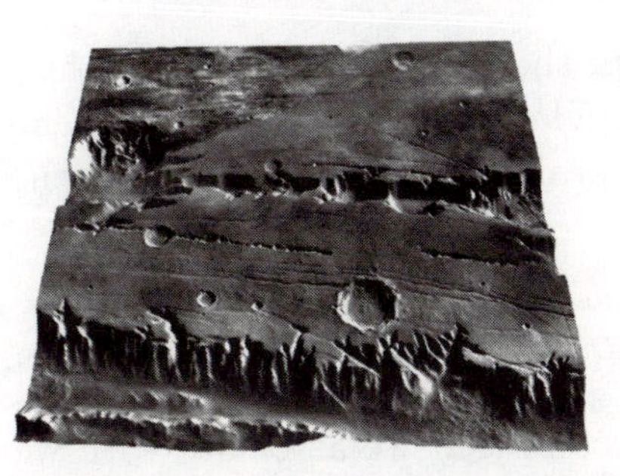

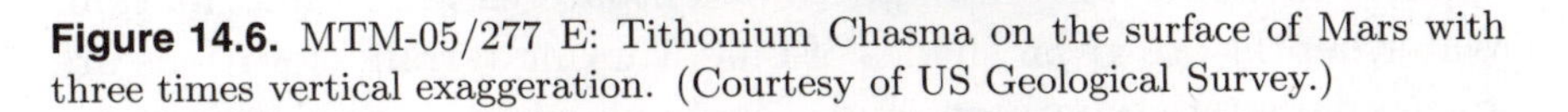

Figure 14.6. MTM-05/277 E: Tithonium Chasma on the surface of Mars with three times vertical exaggeration. (Courtesy of US Geological Survey.)

fects of scale were discussed in Chapter 12 as well, with regard to histograms, bar charts, and log plots.

Some visualization packages give the choice of constant or smooth shading. Figure 14.7 illustrates a pseudo-smooth version of Figure 14.3; by increasing the number of evaluations, the rendering simulates smooth shading.[1] If the phenomenon to be visualized is not

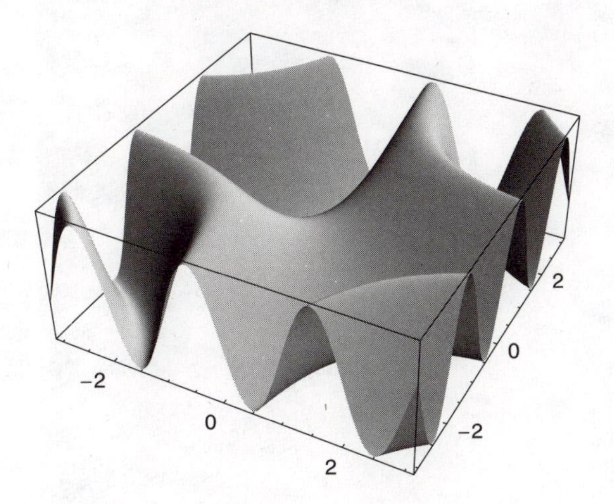

Figure 14.7. A pseudo-smooth rendering of $f(x, y) = \sin(xy)$ over $[-\pi, \pi] \times [-\pi, \pi]$ with 200 evaluations in each axis-direction.

[1] Mathematica 5.2, used here, does not have smooth shading.

a smooth function, then smooth shading might not be appropriate. In particular, triangle meshes that represent empirical data can be difficult to interpret or misleading when smooth shading is applied.

14.2 Color Maps

Often, shades of colors that we associate with water and terrain are applied to digital terrain maps to accentuate elevation information. Water depths are given shades of blue, varying terrain elevations are given characteristic greens, and elevations where it might snow are white. This is called a *shaded relief map* in geographical applications, and an example is illustrated in Figure 14.8. This is a very common form of a *color map.* This technique is also called *false color* or

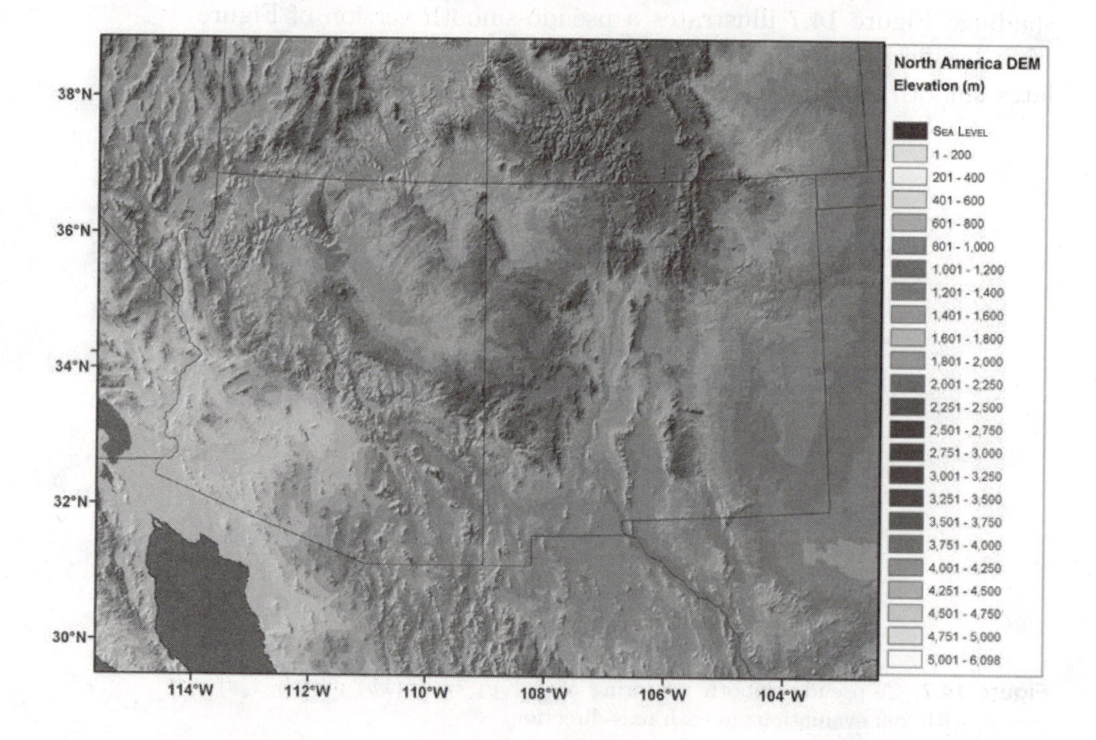

Figure 14.8. A USGS DEM of Arizona and New Mexico plotted in ArcGIS by draping a color map for elevations over a shaded relief. (Courtesy of Ramon Arrowsmith, Arizona State University.)

pseudocolor. For each vertex in the mesh, we look up the appropriate color in a "color look-up table."[2] Alternatively, a photograph of the terrain can be texture mapped onto the surface. (See Section 16.4 for a description of this method.) However, cloud cover can make a clear image of the terrain difficult to obtain, and color maps allow particular features to be accentuated.

Instead of using color to accentuate elevation information, we could use it to communicate precipitation data or landslide risk, as shown in Figure 14.9.[3] This increases the amount of information we can communicate and applies not only apply to meshes, but also to curves and volumes (which are discussed in Section 15.1). Section 13.7 describes how a color map can be constructed for a surface curvature measure and then applied to a 3D mesh.

The simplest form of a color map is a discrete one, which works as follows. First we identify the categories that we would like to differentiate with color. For instance, for Figure 14.8, every 200 meters is given a place in the table as follows.

Color index	0	1	2	...	22	23
Elevation (m)	$\leq$ sea level	1–200	201–400	...	4751–5000	5001+
Color	blue	sea green	yellow-green	...	off-white	white

For each vertex in the mesh, we use its elevation to find the appropriate color in the table. For example, a vertex at 150 m will be assigned sea green. This color has *color index* 1. Elevations are rounded to the nearest meter for use with this table.

Color maps do not have to be constructed as look-up tables; instead, we can construct a function, called a *transfer function*, that continuously maps (transfers) values (e.g., elevation) to a color. In Figure 14.10, we demonstrate this for *reflection lines*. We are given a set of light strips in a plane above the mesh and our eye position. Looking at a vertex **v** on the mesh, we compute the reflection of this incidence vector about the normal at **v**. We find the intersection point, **p**, of the reflection vector and the light plane. Then we find

[2]The graphics hardware will take care of filling in the (triangle or rectangle) mesh element.

[3]Note that full color figures can be viewed at the book's website.

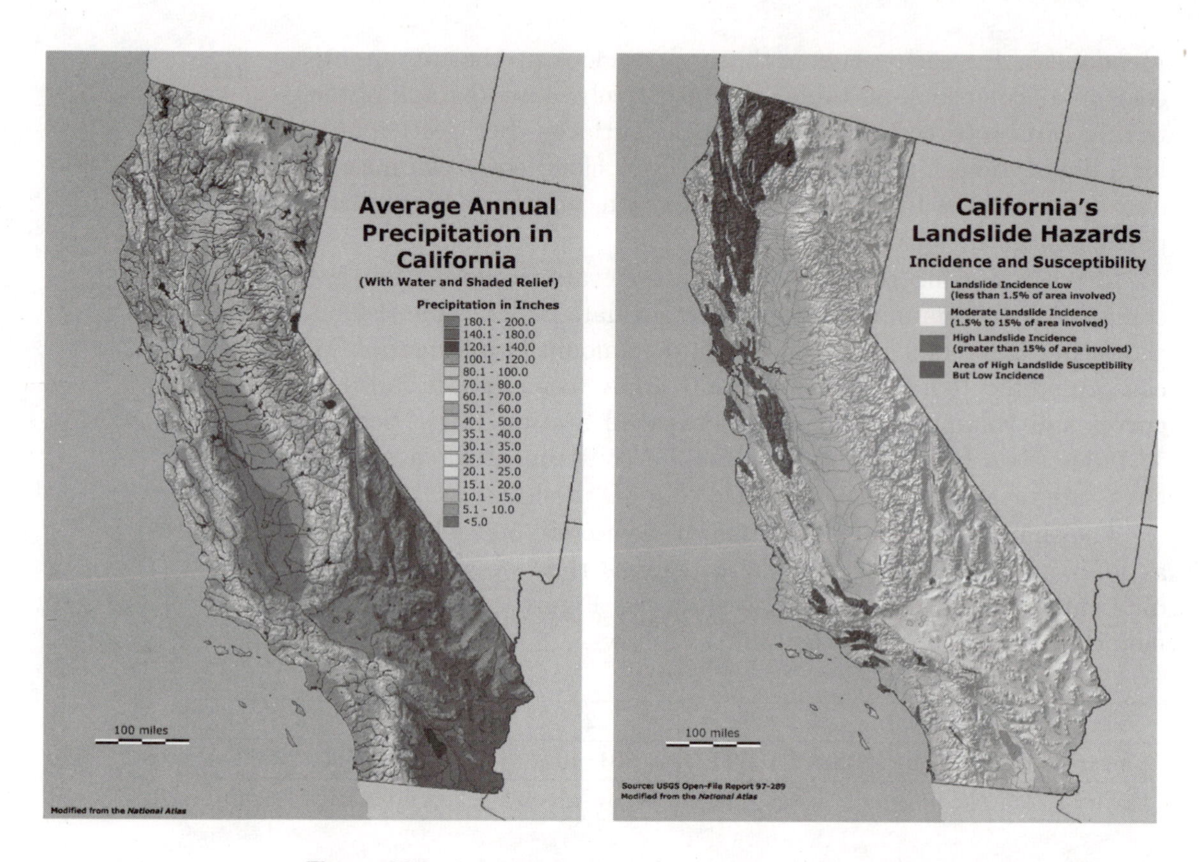

Figure 14.9. A color map that communicates precipitation levels in California (left) and one that classifies areas based on landslide risk (right). (Courtesy of US Geological Survey, Digital Library for Earth Science Education.)

the closest point, $\mathbf{q}$, on a light line to $\mathbf{p}$. The angle α at $\mathbf{v}$, formed by $\mathbf{p}, \mathbf{v}, \mathbf{q}$, is the input of the transfer function, illustrated in the right of Figure 14.10. One can create a different transfer function for each red-green-blue (RGB) color component (see Section 16.1), or a transfer function can be created for transparency. We explore this in Section 15.5.

Commonly, when dealing with categories such as elevation, we construct the color table with linear segmentation. This makes finding the appropriate color index as simple as truncating a linear map. However, the categories don't have to be a linear partitioning; we have seen the benefits of logarithmic scales in Section 12.5.

There are many strategies for choosing and constructing a color map. Depending on the scientific field, certain colors are associated with measurements such as high/low or hot/cold. The color map should speak to the visualization's audience. In the example above, we used the RGB color model to express the elevation categories. More so, though, the hue-saturation-lightness (HSL) model lends itself to creating shades of color. See Section 16.1 for more discussion of these models. An information visualization text is a good place to look for the psychological and cognitive impact of color.

14.3 Contours

An important feature of a terrain map used by hikers is the display of *isolines* of elevation. Isoline is short for isovalue or isopleth line. An isoline indicates where elevation is constant. Another word for isoline is *contour line*, and the process of finding it is referred to as *contouring*.

Figure 14.11 shows isolines are drawn on a map of the area around the south rim of the Grand Canyon. We all know it is wise to stay away from areas where the isolines are very close together because the terrain will be very steep there. We'll revisit the terrain map as a case study in Section 14.4, but first, let's look at contours in more detail.

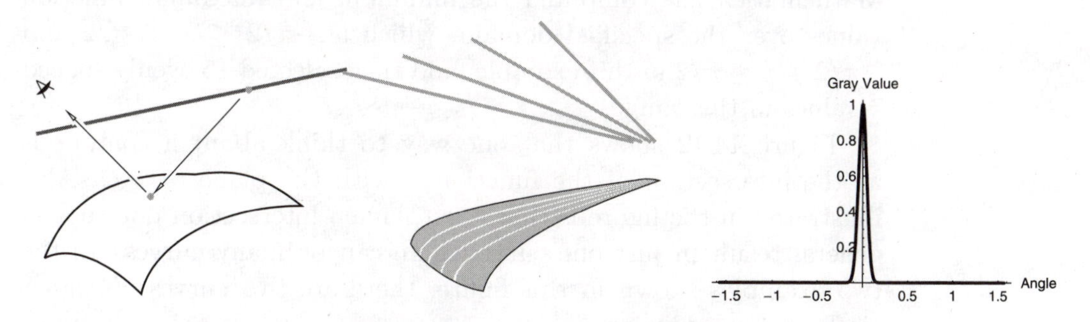

Figure 14.10. Computation of reflection lines (left); reflection lines on a surface with respect to the strip lights above it (middle); and the transfer function $c(\alpha) = \cos^{500}(\alpha)$ for mapping the angle in the reflection line computation to a color (right).

Figure 14.11. Contours on a map of the Grand Canyon. (Courtesy of US Geological Survey.)

The computation of contours is discussed in Section 11.4 for a bivariate function, $f(x, y)$, as the set of all points (x, y) that satisfy the equation

$$f(x, y) = c, \tag{14.1}$$

where the constant c defines the desired isolevel or contour level. Let's continue with the example from Section 11.4, namely $f(x, y) = \sin(x^2 + y^2)$. Figure 14.13 illustrates this function and 15 contours. Mathematica has computed the minimum and maximum function values over the specified domain, which is $-\pi/2 \leq x \leq \pi/2$ and $-\pi/2 \leq y \leq \pi/2$ in this example, and then selected 15 evenly spaced z-values in this range.

Figure 14.12 shows that one way to think about a contour is as the intersection of the function f with the plane $z = c$. Also illustrated in the figure is that one (planar) intersection does not in general result in just one curve; there can be many curves. In the two examples shown in this figure, there are two curves, an inner circle and an outer circle.

Looking carefully at the contour in Figure 14.13, we see that near the borders, the contours are not correct—they all need to be parts of exact circles! Also, the center-most contour circle appears

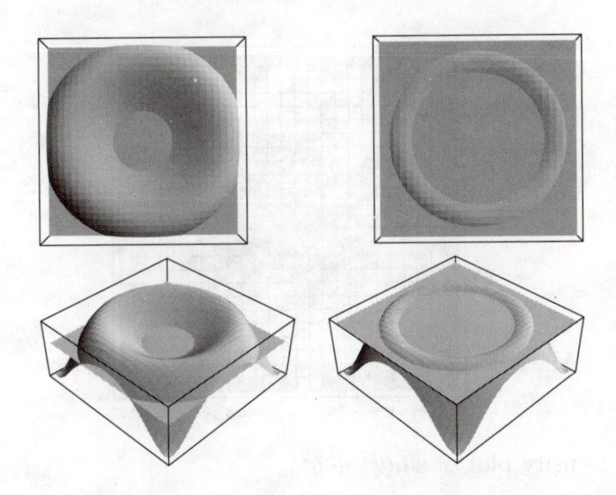

Figure 14.12. Intersection of the plane $z = 0.25$ with $\sin(x^2 + y^2)$ (left). The sine function is plotted on a coarse grid to accentuate the contour's appearance. Intersection with the plane $z = 0.9$ (right). In both examples, two contours are produced.

to be a diamond. Let's look at these problems by first studying how contours are computed.

Consider Figure 14.14, a so-called *density plot.* It shows a 2D array with vertices (x_i, y_j); these vertices correspond to the evaluation points illustrated in Figure 14.13. Four evaluation vertices form a

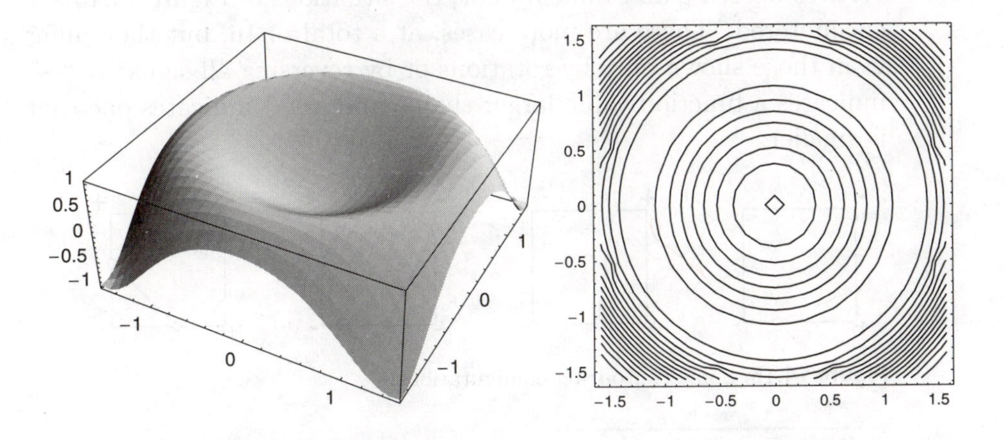

Figure 14.13. Plot of the function $\sin(x^2 + y^2)$ over $[-\pi/2, \pi/2] \times [-\pi/2, \pi/2]$ evaluated on a 25 × 25 grid (left) and contour levels at 15 z-values (right).

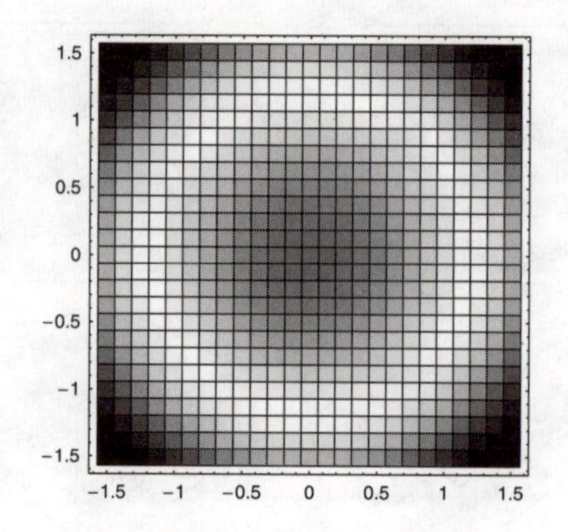

Figure 14.14. Density plot of $\sin(x^2 + y^2)$.

cell. A color map is created that maps the range of function values onto gray values in the range $[0, 1]$. Black (gray value 0) corresponds to the minimum z-value,and white (gray value 1) corresponds to the maximum z-value. Each cell is colored according to the function value at its center.[4]

The cells are used to compute a contour line as follows. For each cell, we check if its four vertices have function values larger or smaller than c. If all four are smaller than c, then the cell is assumed to be below the contour level, and no output is generated. If all four are larger than c, again there will be no output. In cases where some cell vertices exceed c and some do not, the situations of Figure 14.15 are encountered. There are more cases, at a total of 16, but they differ from those shown only by rotations or by reversing all signs. A "+" indicates a function value larger than c and a "–" indicates one that is smaller.

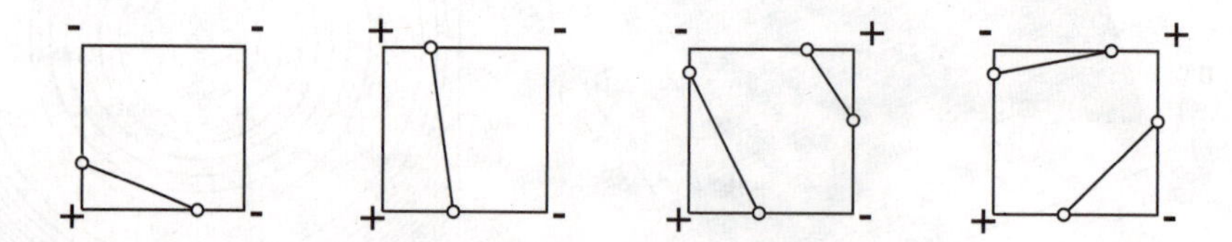

Figure 14.15. Some contouring configurations.

[4]If our given data at the cell vertices are not from a known function, but rather from discrete data, then we would use *bilinear interpolation* to approximate an average value for the cell. See Section 11.2 for details.

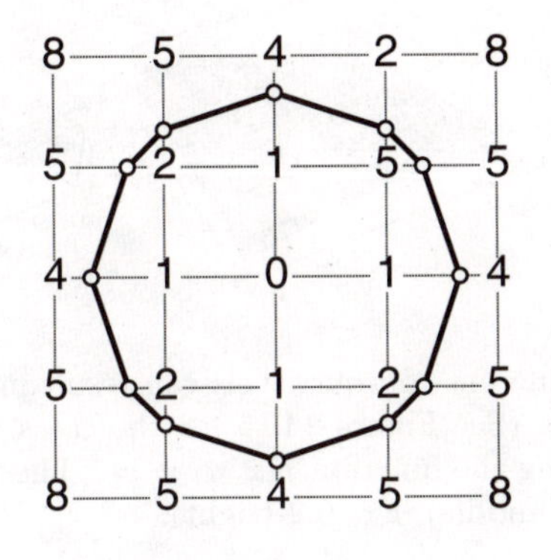

Figure 14.16. A simple contour example of the function $f(x, y) = x^2 + y^2$ over $[-2, 2] \times [-2, 2]$ for contour level $z = 3$.

Suppose there is a sign change on the ("horizontal") edge (x_i, y_j), (x_{i+1}, y_j), as depicted on the bottom edge in the left-most diagram of Figure 14.15. Using a root finding method such as Newton-Ralphson from Sections 10.2 and 11.5, we could find the "exact" point on this edge where $f(\hat{x}, y_j) = c$; however, if the grid is dense, the calculation would be very slow.[5] Instead, we assume that the grid density is sufficiently dense, and we make the simplifying assumption that on each cell edge, the function f behaves linearly. By finding where in the domain the linear interpolant (see Section 3.4) satisfies $f = c$,

$$t = \frac{c - f_{i,j}}{f_{i+1,j} - f_{i,j}},$$

the value $\hat{x}$ can be quickly found to be

$$\hat{x} = x_{i,j} + t(x_{i+1,j} - x_{i,j}).$$

A small example of a computed contour is shown in Figure 14.16.

Of the configurations in Figure 14.15, the first two are trivial: we connect the shown vertices and move on to another cell. The last two configurations are trickier; either of the shown solutions (or even a third) might be correct, as illustrated in Figure 14.17. These are

[5]Root-finding methods find the point where the function intersects the $z = 0$ plane. We can simply translate our function by $-c$ to modify our problem for these methods.

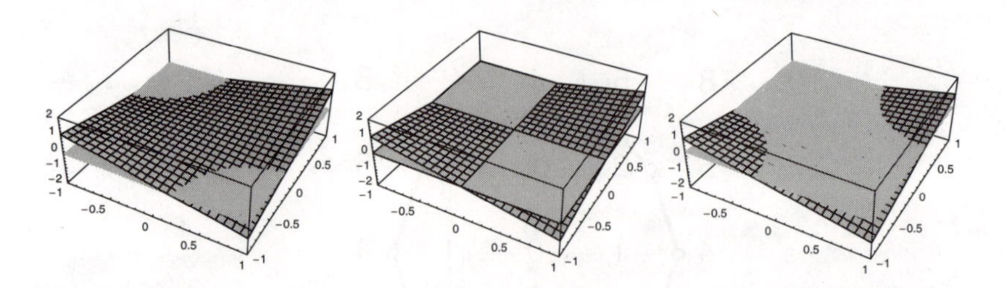

Figure 14.17. Three different contour levels can result in the same cell case but different connectivities. (See Figure 14.15 for the cases.) The cell, represented as a plane, is intersecting the function $f(x, y) = xy$. The three contour levels are $z = -0.2$ (left); $z = 0$ (middle); $z = 0.2$ (right).

called *ambiguous cases.* Without knowing the underlying function, there is no way of knowing the correct connectivity; one approach to making a better decision is to look at neighboring cells, although this is not foolproof. Figure 14.18 illustrates how this choice can influence the final contour. If the function is known, as in the example above, then the function can be sampled within the cell to resolve ambiguous cases.

In the visualization community, this method of constructing a contour is called *marching squares* because the algorithm moves through all the cells and calculates the contour elements. The 16 cases, derived from the registration of a sign at the four vertices of a cell, can be encoded with four bits. This allows for fast access to the connectivity rules. The marching squares paradigm works well for rendering; however, for some applications, the connectivity of the entire contour is required. In such a case, it is necessary to track the

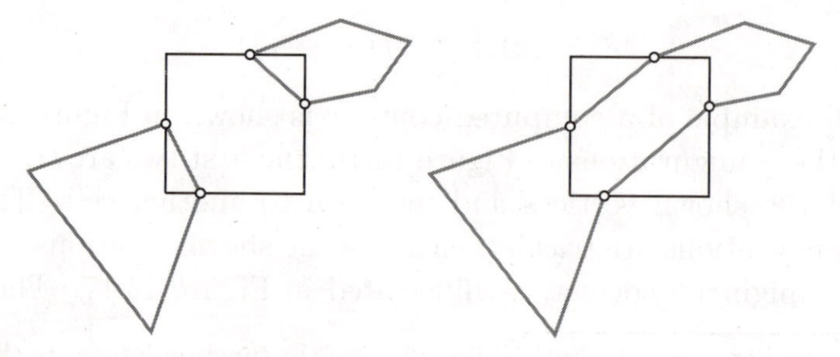

Figure 14.18. Ambiguous cases for connectivity in one cell can lead to incorrect contour results.

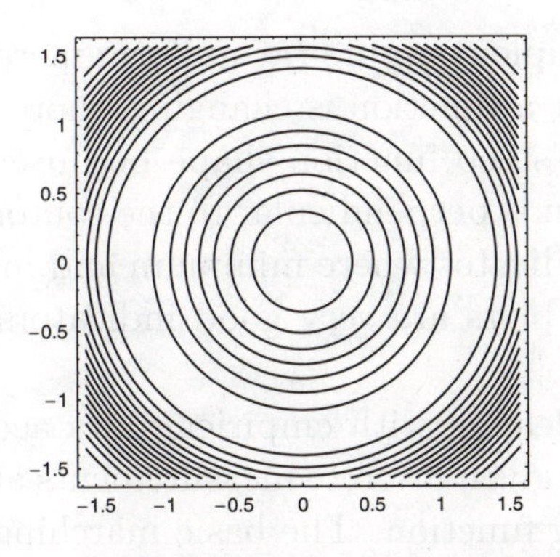

Figure 14.19. An improvement over the contour from Figure 14.13 by computing the contour over a denser grid. The function is evaluated on a 90×90 grid.

contour. We look at marching cubes for scalar data over volumetric (3D) data in Section 15.2.

Looking at the marching squares method for contouring, we can see that the accuracy of a contour requires a sufficiently dense mesh so that the piecewise linear approximation over each cell is reasonably accurate. Thus, we can improve on the contour of Figure 14.13 by increasing the cell density, as illustrated in Figure 14.19.

Figure 14.20 illustrates a combination of the contour plot in Figure 14.19 and the density plot in Figure 14.14. This is called a *shaded*

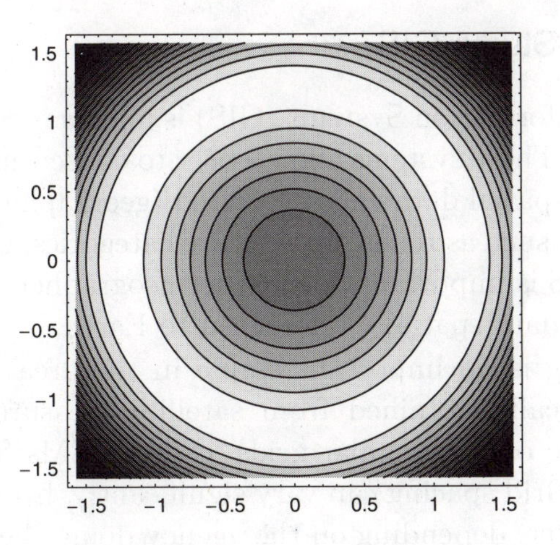

Figure 14.20. Contour lines of $\sin(x^2 + y^2)$ combined with a density plot.

contour plot. Multiple contours (at uniform increments of z) give us an idea of how fast a function is changing; more closely spaced contour lines indicate steep function shape because the gradient of the underlying function is perpendicular to the contour line. In contrast, the density plot indicates where minimum and maximum values are. Together, the two plots are very good indicators of the behavior of the data.

When we are dealing with empirical data acquired from a scanning device, we are given discrete measurements at the vertices rather than an underlying function. The basic marching squares algorithm works similarly; however, we no longer can sample the function arbitrarily to resolve ambiguous cases. One possibility is to locally fit (piecewise) polynomials to the given data and then contour the approximating function.

Some phenomena, such as weather, are impossible to measure on a regular grid. Instead, measuring stations are scattered. As discussed in Section 13.5 a triangle mesh is a tool for creating connectivity between scattered data. From watching the nightly news, we are all familiar with isotherm (constant temperature), isobar (constant pressure), and isotach (constant windspeed) maps. Contours help us track changes in weather conditions, and in turn predict special weather events such as hurricanes.

14.4 Case Study: GIS

Geographic Information Systems (GIS) is informatics applied to geographic data. These systems allow a user to collect, manage, analyze, and visualize spatial data and associated (geographically referenced) attribute data such as roads, agriculture categories, or rainfall statistics. Thus GIS is important not just for geographers, but for anyone working with data spatially referenced to Earth.

Contouring is an important staple in the area of GIS. Terrain data are typically obtained from satellite measurements and are made available on rectangular grids, called DEMs for digital elevation models. Grid spacing can vary significantly, from several meters to one kilometer, depending on the agency doing the measuring, the area being measured, and what is being measured. In this method,

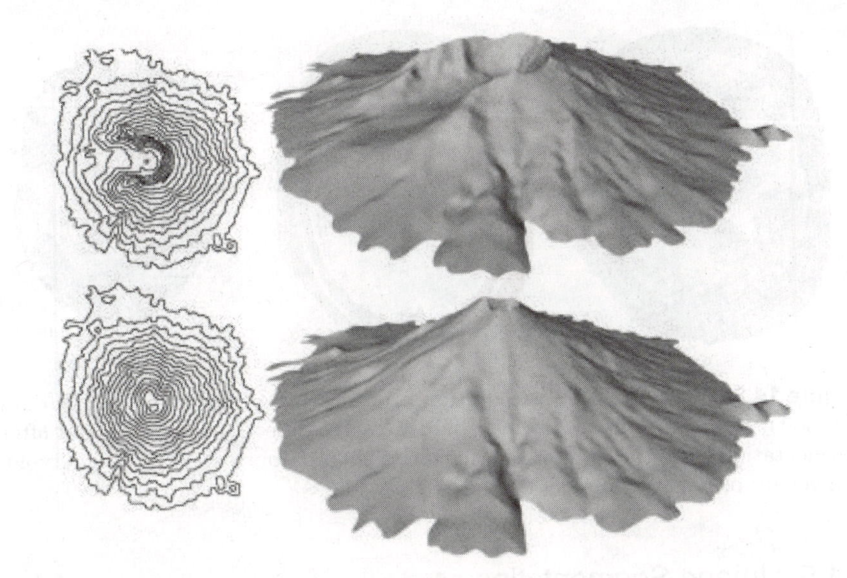

Figure 14.21. Contour maps and 3D models of Mount St. Helens: After the 1980 eruption (top) and before the eruption (bottom). (Courtesy of J. Schneider, Technical University Munich, Germany.)

an elevation $f_{i,j}$ is given at each vertex of a cell. Two such contour maps are shown in Figure 14.21 together with 3D models of the corresponding terrains.

When computing with terrain models, DEMs are often converted to triangle meshes. This is simply done by splitting each quadrilateral cell into two triangles by adding a diagonal. The resulting triangle mesh is contoured very easily since the intersection of a triangle and a plane is a trivial task. However, the ambiguous cases from the marching squares algorithm above are not resolved by using a triangulation process because the choice of diagonals in a quadrilateral is arbitrary. Thus contouring errors are still possible.

Today's widespread use of GIS has been made possible by the Global Positioning System (GPS), which is a system of satellites and computational systems. Two companies, ESRI and MapInfo, dominate the commercial GIS market, although several open source packages, such as GRASS GIS, are available. Google Earth is a tool that allows everyone to work with GIS.

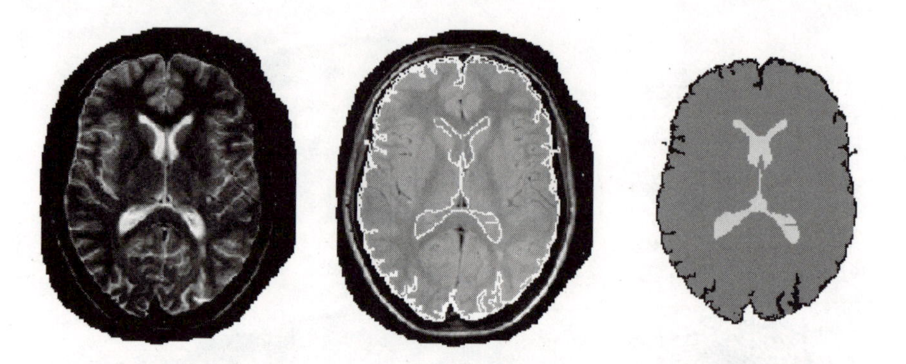

Figure 14.22. Brain image segmentation process: slice of the original MRI data set (left), after contours are formed by using edge detection (middle), and after segmentation into gray and white matter (right). (Courtesy of A. Lundervold, University of Bergen, Norway.)

14.5 Image Segmentation

Another application of data generated on a rectangular mesh is that of medical imaging. Techniques such as MRI, PET, or ultrasound yield volumetric data[6] sets that are typically analyzed slice by slice.

We may view an image as a discrete bivariate function, $f(x_i, y_j)$, in which the function value for each pixel (x_i, y_j) is a gray value. If we contour at a certain value, we find all pixels with the desired value. *Edge detection* aims at finding contours in the image, which flag rapid changes in gray values; this task is more difficult because the contour level itself is not known.

Figure 14.22 shows an example of the segmentation process showing the original MRI data set, after contours are found using edge detection, and after segmentation of the data into gray and white matter. Cerebrospinal fluid (CSF) is shown at the center of the right image.

A rapid transition in pixel intensities is characterized by a high gradient $\nabla f(x_i, y_j)$. This is somewhat costly to compute, and hence a typical edge detector employs a *filter*. An example is the Prewitt filter, which finds large intensity changes in the x- or y-directions. At a pixel (x_i, y_j), the partial derivative of f in the x-direction is

[6]We look at volume data sets in detail in Chapter 15.

approximated by[7]

$$f_x(x_i, y_j) = f(x_{i+1}, y_i) - f(x_{i-1}, y_i).$$

In the same way, we may compute the x-partials at (x_i, y_{j+1}) and (x_i, y_{j-1}). Taken together, we may express the averaged x-partial, $P_x(x_i, y_j)$, by the *mask*:

$$\begin{array}{ccc} -1 & 0 & 1 \\ -1 & 0 & 1. \\ -1 & 0 & 1 \end{array}$$

This is pictorial for

$$\begin{aligned} P_x(x_i, y_j) = {} & f(x_{i+1}, y_{j+1}) + f(x_{i+1}, y_j) + f(x_{i+1}, y_{j-1}) \\ & - f(x_{i-1}, y_{j+1}) - f(x_{i-1}, y_j) - f(x_{i-1}, y_{j-1}). \end{aligned}$$

In regions where there is no change in the x-direction, this filter will return 0, and it will return high values where rapid changes take place. Similarly, we define $P_y(x_i, y_j)$ by the mask

$$\begin{array}{ccc} 1 & 1 & 1 \\ 0 & 0 & 0. \\ -1 & -1 & -1 \end{array}$$

Figure 14.23. Edge detection. A filter is applied to detect a high gradient in the discrete bivariate function (the image). Zooming in on the capital D, we see that the gradient is mapped to a color map with shades of gray.

[7]There should be a division by the pixel spacing, but since that is the same for all pixels, it is safely ignored.

The gradient at (x_i, y_j) may now be approximated by (P_x, P_y)—its magnitude will flag rapid intensity changes in an image. Thus, in Figure 14.23, only the pixels with a gradient higher than a certain threshold are left after application of the filter. To achieve smooth edges, the color map has shades of gray rather than just black and white.

More involved filters exist as well as other kinds of edge detection methods known by names such as *active contours* or *snakes*; we will not cover those here.

14.6 Problems and Experiments

1. Experiment with the effects of scaling a height map.

2. Experiment with color maps that are based on a function's gradient. What other properties of a function might be associated with a color map so that more information about it is revealed?

3. Choosing a reasonable number of contours is an important step in revealing shape information. But choosing too many contours can be confusing. Experiment with algorithms for a function-dependent method for estimating the number of contours.

4. More difficult than determining the number of contours is determining the grid density. Experiment with algorithms for determining this automatically.

5. Image segmentation: Using software such as Mathematica, compute the gradient over a function of your choice. (You will need to choose a domain and evaluation partition.) Use a color map to identify points where the gradient's magnitude is large.

15

Volume Visualization

Volume visualization refers to the visualization of functional phenomena in 3D space. At positions in 3D, we are given scalar values, vectors, or tensors (higher dimensional data). The primary focus of this chapter is on two techniques, contouring and direct volume rendering, for visualizing a scalar-valued function defined over a volume. Some attention is given to visualizing vector fields and tensors (higher dimensional phenomena). Finally, we bring attention to the relatively new field of haptic visualization.

15.1 Scalar Data over a Volume

When the given data are scalar values over a volume, the domain typically falls into one of the following three categories:

- a rectilinear grid of voxels,[1]
- a rectilinear grid of cells, or
- a set of scattered 3D points with a connectivity formed by tetrahedra.

Figure 15.1 illustrates these three domains.

[1]The term is short for "volume element."

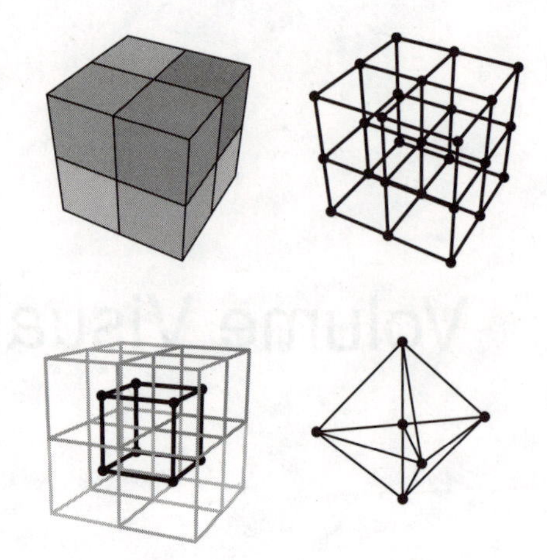

Figure 15.1. Domains for the given data for volume visualization: a rectilinear grid of voxels where each cube has one value associated with it (top left); a rectilinear grid of cells where each vertex has a value (top right); a cell defined from eight voxels (bottom left); and scattered 3D data and their tetrahedra (bottom right).

A *voxel* is said to be the 3D version of a *pixel*.[2] If we partition the space over which discrete samples are given on a rectilinear grid, then volumes in the shape of cubes, or voxels, are the result. Analogous to the pixel, and as illustrated in Figure 15.1, a voxel has its center sample value defined over it. However, algorithms based on voxels rarely use them this way; it is assumed that the value is changing throughout the voxel, and it is approximated by examining neighboring voxels.

A *cell* is a cube as well, but at each of its eight vertices, a value is defined. Thus the values within a cell can vary. As we will see in the sections to follow, in some applications, the cells are created from the voxels. This idea is illustrated in Figure 15.1 as well. However, a rectilinar grid of cells can be generated without voxels, such as in a simulation where we sample (evaluate) a function.

Tetrahedral meshes to scattered 3D data points are the 3D analogy of 2D triangle meshes (in the plane), which were discussed in

[2]This generalization is not really very accurate, however. A pixel is an image space element that can take one value only, but a voxel is an object space element with (possibly) varying value.

Section 13.5. A tetrahedron is formed from four noncoplanar points. Issues with respect to creating tetrahedral meshes, data structures, and neighbor information are very similar to those of triangle meshes.

As Figure 15.1 illustrates, we assume that the rectilinear domains are isotropic, which means the same spacing occurs in each axis direction. Further, we assume they are axis aligned. The domain could be anisotropic (arbitrary spacings), curvilinear, or—for scattered data—use elements other than tetrahedra. These other choices tend to complicate the algorithms, so here we stay with the basic and more commonly encountered situations. If the data are not given on the partition needed for a particular algorithm, sampling and interpolation might take place to create a new partition. This topic is discussed in Section 16.5.

Let's look at three examples of applications in which we are given scalar values over voxels, over cells, and over scattered data points.

Example 1. Voxels from an imaging application. Medical applications certainly have been a driving force behind research in volume visualization. CT scans produce approximately 64 slices of data, from one to five millimeters apart. Typically, each slice will have a resolution of $512 \times 512 \times 12$ bits.[3] The 12 bits hold *radiodensities*, which are assigned according to the materials encountered, such as bone, tissue, or organ. It is also possible to assign multiple material types and their percentages to a voxel. Figure 15.2 illustrates a rendering of slices. Section 14.5 on image segmentation showed that thresholds for radiodensities are selected and paired with a transfer function. (We revisit this transfer function in Section 15.5 below.)

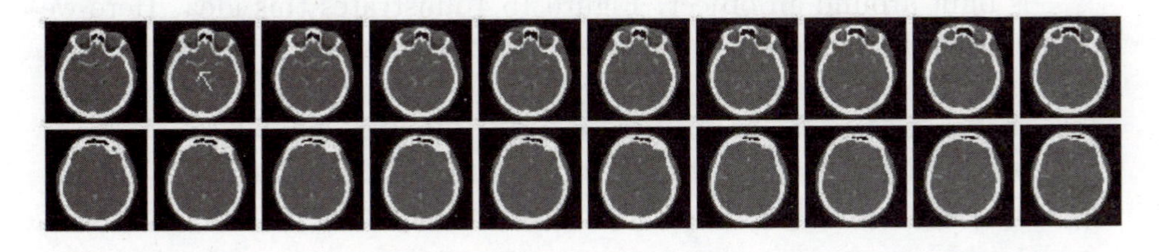

Figure 15.2. Tomograms: 2D slices produced by a CT scan. Bone, corresponding to high radiodensity, is mapped to white. Blood vessels are visible due to a contrast agent injected into the patient.

[3] Of course, the resolution figures will change with technology.

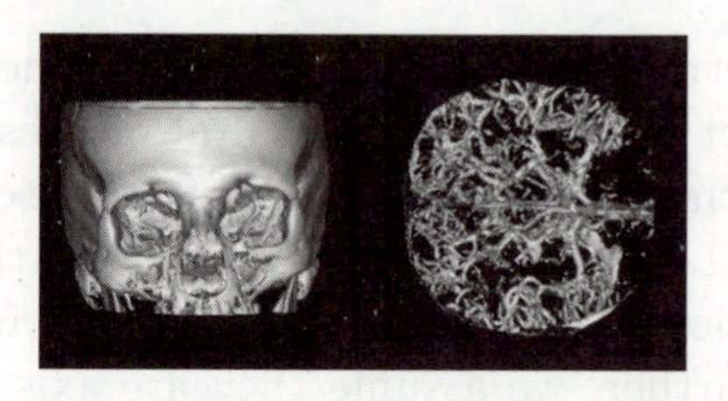

Figure 15.3. Volume visualization of bone and blood vessels.

If we consider the slices as part of a whole, we obtain the rectilinear voxel structure. Figure 15.3 illustrates two volume visualizations of the same data set.

Example 2. Cells from a computational science application. Fluid flow analysis is extremely important for many applications in engineering and the natural sciences. For example, analysis of gaseous and liquid flow allows us to build more efficient cars, spacecraft, turbine engines, and ships. Because of its complexity, visualization of flow is a necessity for a meaningful analysis and verification of theories.

Before the 1960's, the only way to "visualize" flow was to build models and physically observe flow behavior. Once computers became powerful enough to handle the large and complex calculations needed to simulate flow, the field of computational fluid dynamics (CFD) emerged. Numerical flow simulation consists of three phases: grid generation, flow calculation, and visualization. A grid structure is built around an object. Figure 15.4 illustrates this idea. Here we are discussing volume visualization, so we assume it is a 3D object, such as a spacecraft. Many types of grid structures can be used, such as rectilinear or curvilinear.[4] For the flow simulation, a system of equations based on the Navier-Stokes or Euler equations is solved. The solution will result in quantities such as momentum, density, or

[4]Rectilinear and curvilinear grids are examples of structured grids whose elements are topologically equivalent to a cube. Curvilinear grids have straight lines for edges. Unstructured grids are formed by tetrahedral meshes. The difficulty in computing over these elements increases with the order they are given. Many hybrid grid structures exist as well.

Figure 15.4. Flow phenomena about a delta wing. Flow behavior is studied using streakline particle traces (gray) and vortex cores (white), which were computed by the Unsteady Flow Analysis Toolkit (UFAT) developed by NASA. (Courtesy of David Kao (NASA); Delta Wing courtesy of Neal Chaderjian (NASA).)

stagnation energy. From these quantities, velocity can be derived. Finite element methods are used to solve the equations at either the grid vertices or cell centers. Finally, the cells and scalar or vector values are ready to be visualized. Figure 15.4 illustrates flow phenomena computed from a curvilinear grid due to the complexity of the geometry.[5]

Computationally, real-life CFD problems are a huge challenge. Lane [13] gives an example of computation for a Harrier jet. The jet's grid might take 45 MB of disk space, the solution data might take 56 MB, resulting in 101 MB of disk space for one time step. Scientists need to analyze many time steps—for example—90,000 steps to study descent and landing. Interactive visualization is also a challenge, and visualization researchers are working on ways handle these massive amounts of data. Computational scientists are always pushing the limits of computing power, memory, and novel algorithms for managing the data.

[5]We encounter some elements of this figure in Section 15.9.

Example 3. Tetrahedral meshes from data acquisition technology. Many fields in engineering and science have developed data acquisition methods for capturing data in a volume. For instance, seismic imaging technologies are used to allow geoscientists to build subsurface models for predicting oil and gas reservoir size and shape. Acoustic engineers take sound readings in a room for the purpose of modeling and visualizing the propagation of sound caused by the walls, furniture, and other artifacts.

These three examples are representative only of volume data sources and applications. We could name many more; for instance, the following are common:

- Additional medical scanning techniques: MRI, fMRI, and PET.
- Nondestructive inspection of composite materials or mechanical parts using CT.
- Exploring biological structure and function with confocal microscopy.
- Observing ribosomes with electron microscopy.
- Bringing to life the dust and gas found throughout the galaxy from data acquired by the Hubble telescope. (See wonderful visualizations at http://vis.sdsc.edu/research/orion.html)

A fascinating observation is the variation of scale for which volume visualization takes place. We see problems at the micro- and macro-level. Simultaneous visualization of phenomena at these scales is an open research problem. Additionally, some data sets are "large volume data sets"; this term is relative and, of course, will change as computing power and memory change. In 2007, a very large data set is considered to be greater than 100 GB.

In this chapter, we assume that the data are ready to use. In many real-world applications this is not the case, and data processing in the form of resizing, sampling (interpolation), rezoning, restructuring, and editing of unwanted data is necessary. These issues are addressed in Section 16.5.

15.2 Contouring

In this section, we look at the visualization of scalar values over a rectilinear grid of cells with contouring. Contouring is a type interrogation. It provides a means of visualizing lower-dimensional representations for the purpose of understanding the more complex, whole form. In Section 14.3, we looked at the computation of piecewise linear approximations to contours of bivariate functions with the "marching squares" method. Here we increase the dimension, and thus look at *contours* for functions $f(x, y, z)$, as the set of all points (x, y, z) that satisfy the equation

$$f(x, y, z) = c, \tag{15.1}$$

where the constant c defines the desired *isosurface*. The algorithm we use here to approximate the contours is called *marching cubes.* The cubes are the cells from Section 15.1. The cells might be derived from voxels, as illustrated in the bottom left of Figure 15.1, or the cells might be generated for evaluating an empirical model.

Let's mirror the discussion from Section 14.3 and first work with a function with which we are familiar, namely, $f(x, y, z) = x^2+y^2+z^2$. Figure 15.5 illustrates two contours

$$f(x, y, z) = 0.5 \quad \text{and} \quad f(x, y, z) = 1.5.$$

In this figure, the contour level $f = 1.5$ is clipped by the viewing volume so we can see the sphere $f = 0.5$ inside; the outer sphere is really a complete sphere. Mathematica has added 2D contours of constant x-, y-, and z-values to the surface to help the eye see the shape. This is a contour of the 3D contour! With this figure, we see that contouring is a method for *surface extraction.* Once the surface is extracted, it can be rendered with the normal graphics pipeline.

Let's look at some details of the marching cubes method. First, we are given a rectilinear grid of cells. In Figure 15.5, we have a 10×10 rectilinear grid defined over $[-1, 1]$ in each dimension. The function f is then defined at each vertex of the rectilinear grid. For a given contour level, we "march" through each cell, and check the function values at the eight vertices. If all eight are above or all eight are below the contour level, then no more processing of this

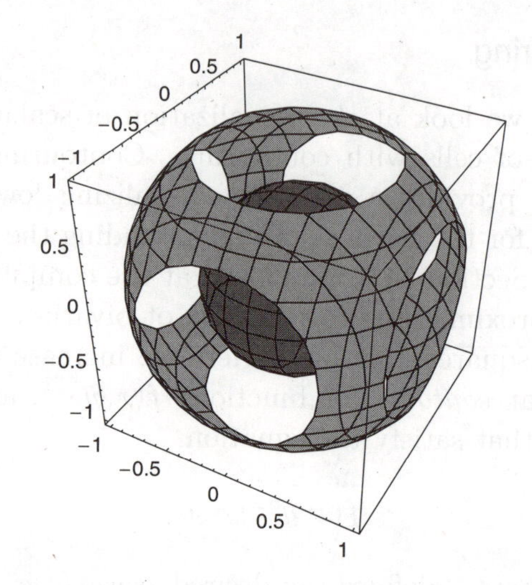

Figure 15.5. Contours of the function $f(x, y, z) = x^2 + y^2 + z^2$. The contour $f = 0.5$ is the inner sphere, and $f = 1.5$ is the outer sphere, clipped by the viewing volume.

cell is necessary. If only some vertices are on or above the contour level, then we process this cell. Figure 15.6 illustrates constructions for four different cases; piecewise planar (triangle) elements are constructed to approximate the contour. We make the simplifying assumption that the function behaves linearly over the cell's edges.

Considering all configurations of vertices and contour level, we find there are $2^8 = 256$ cases. However, if we exclude those that are identical through reflection or symmetry, then there are 15 entries

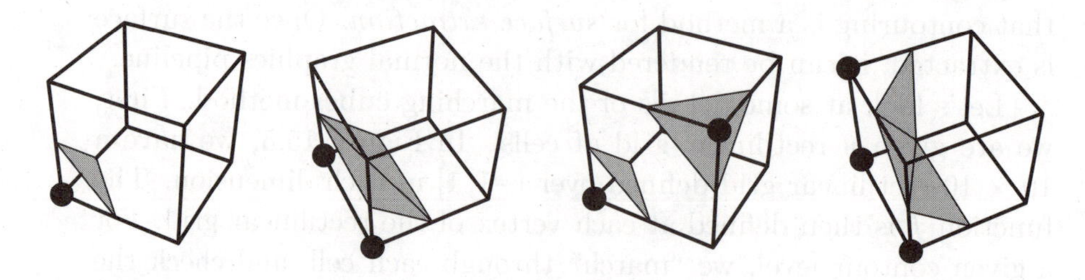

Figure 15.6. Some marching cubes cases. The black dot at a cell vertex indicates the function value there is on or above the contour value.

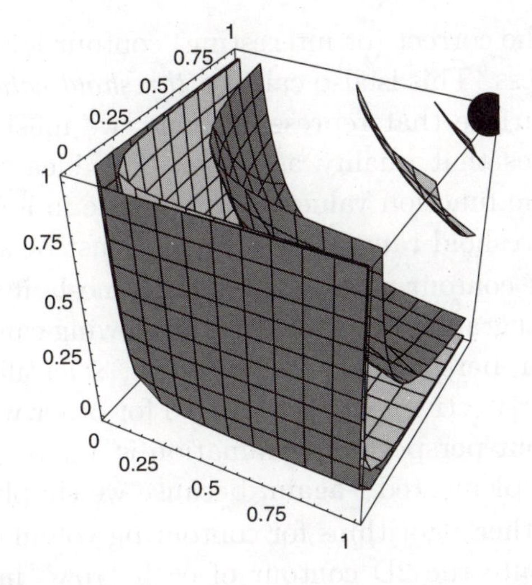

Figure 15.7. Several contours of the trilinear function $f(x, y, z) = xyz$ over $[0, 1]$. The cube vertex with the black dot has value one; the remaining seven vertices have value zero.

in the case table that documents the topological states. Figure 15.7 illustrates some contours of a trilinear function, and it illustrates that the triangle elements are, most certainly, just a rough approximation of the true shape within the cell. (This figure corresponds to the left-most case in Figure 15.6.) Except for special cases, contours are closed. By simply "marching" from cell to cell, holes in the contour can be created. To avoid this, in some cases it is necessary to consider neighboring cells. One remedy is to add some additional "complementary cases," which are simply different configurations of triangles over a cell. These have been designed to knit together a closed contour. If the function over the rectilinear grid is known, as it is in our case, then the function can be sampled within the cell to clarify any ambiguities.

Instead of dealing with the problems of marching cubes, it is also possible to create two tetrahedra from each cube. This solution implicitly resolves ambiguities, correctly or not. Tetrahedra will result in many more triangles defining the contour. (Decimation methods, discussed in Section 16.5 can be used to reduce the number of triangles.) If the given data were scattered to begin with, then we have no choice but to work with this structure.

Choosing the correct (or interesting) contour level can be a trial-and-error process. This is also called a *threshold value*. For instance, to extract a surface that represents bone, we must choose a range of radiodensities that qualify as bone. The idea that we contour precisely for one function value is not realistic in a computing environment. A threshold range, $[c - \epsilon, c + \epsilon]$, must be specified.

Because the contour process results in a mesh, it can be rendered with the normal graphics pipeline. Thus, viewing can easily be either orthographic or perspective. Of course, scientific value can be distorted by perspective; see Section 16.3 for other ways to achieve a 3D effect without perspective. Animation is a good solution as well and is easily implemented—again, because we simply have a mesh.

There are other algorithms for contouring volumes. For instance, we could compute the 2D contour of each "row" in the rectilinear grid, and then connect the planar contours. This method, called *multiplanar reconstruction*, was used frequently before memory and computing power could handle the large real-life volumes produced. It is still used for some special visualizations.

15.3 Case Study: Health Care

K. Frenkel [5] tells a riveting story from the 1980s of a young man who suffered a crushed pelvis in a car accident. His orthopedist determined that the fracture was too complicated to operate on and elected to treat him with a few months of traction. This would have left the patient permanently crippled. Luckily, this young man's father knew of research in CT scan volume rendering. He sent his son's scans to researchers in this area, and they were able to create the shattered pelvis from any angle. With this new information, the surgeon could see the extent of the fracture and locate the key fragments. The pelvis was operable after all, and the surgeon was able to plan and execute the surgery. Three months later, the young man had a full-range of hip motion.

This story is some years old, but the scenario is still repeated today as volume rendering is just gathering momentum in the medical community. Approximately 3–5% of US hospitals and radiology practices use volume rendering. University-level medical centers are the largest practitioners. For many years, medical practitioners were concerned about misdiagnosis due to computer-generated artifacts and pseudo-color. Another hinderance to the growth of volume visualization has been competing scanning vendors and their proprietary data formats.

For many years, radiologists have looked at films, called *tomograms*, which depict the individual slices from a scan. Radiologists then mentally put the slices together and identified irregularities. A growing problem has been the availability of radiologists. In 2002, the *American Journal of Radiology* reported that the deficit of radiologists was at 5% and could grow to as high as 50% by 2020. Volume visualization will reduce the time a radiologist needs to analyze a patient's scan. Additionally, the information in the images can be easily communicated to nonradiology physicians and laymen (such as in a courtroom). To satisfy the growing need for automatic diagnostics, efforts have been under way for automatically identifying irregularities.

15.4 Direct Volume Rendering

In this section we look at visualization of scalar values over a rectilinear grid of voxels with *direct volume rendering* (DVR). This class of method maps (3D) voxel data onto a 2D image (screen) space. No geometry is created, just a 2D image. DVR methods are particularly useful for amorphous features such as gas, clouds, or fluids. The negative aspect, however, is that the entire voxel space must be traversed for a new view.

There are many approaches to DVR, and new approaches are still being developed. To understand the general idea behind DVR, let's look at a simplified version of one of the most basic approaches, called *volume ray casting.*

With four diagrams, Figure 15.8 illustrates the basics of volume ray casting. In diagram 1, a ray is cast from a pixel in the image through the volume. This ray traverses several voxels; the task at hand is to accumulate the scalar values encountered along the way, as only one value can be recorded at the pixel.

Diagram 2 illustrates that sampling is the next step. Here we have five uniformly sampled points. We must determine function

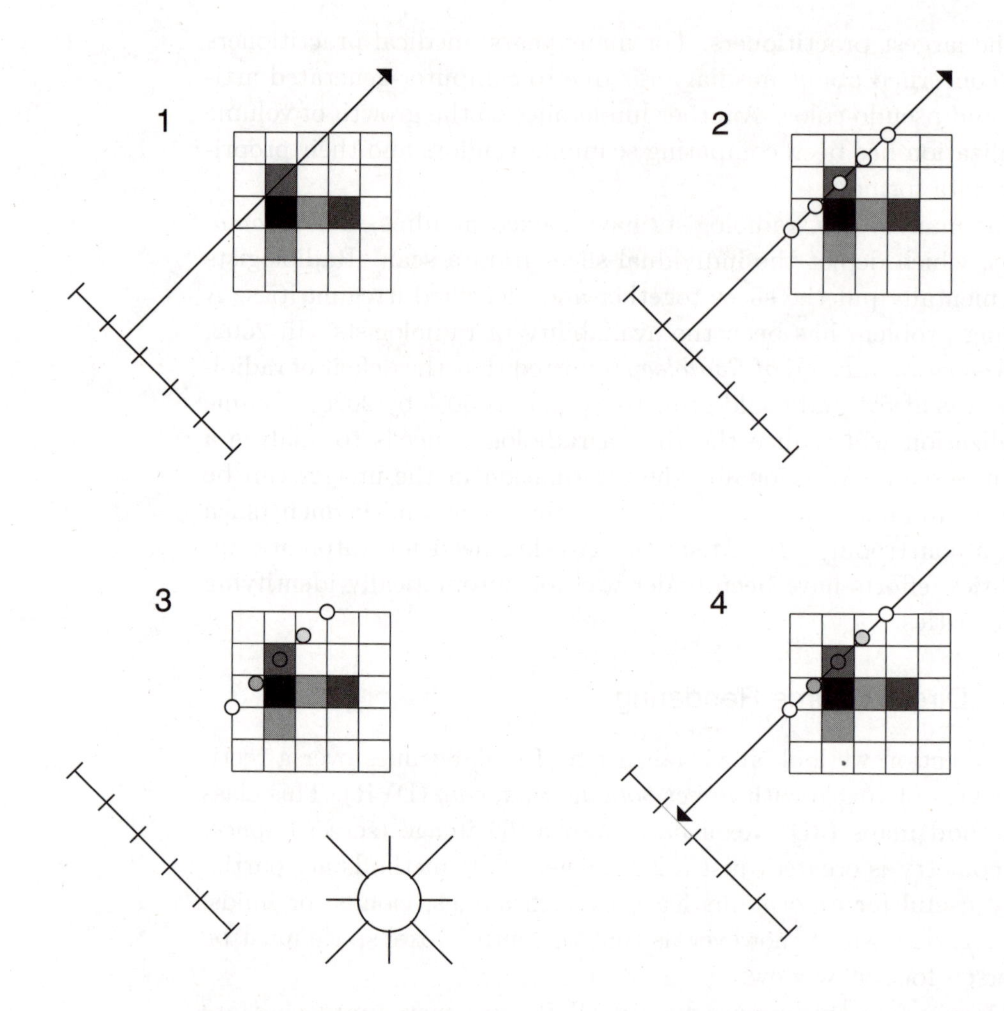

Figure 15.8. Ray casting depicted as four steps. Diagram 1: Cast a ray from a pixel. Diagram 2: Sample along the ray and interpolate to voxel values. Diagram 3: Using gradients at the sample locations, apply lighting. Diagram 4: Composite the sample results and color the pixel. (Figure concept from Wikipedia's volume ray-casting page.)

values at these samples. It is unlikely that a sample point will be precisely at the center of a voxel. To find a good value, some sort of interpolation is in order. A nearest-neighbor function is one possibility: choose the function value of the voxel in which the sample resides. A better solution, and also quite simple, is trilinear interpo-

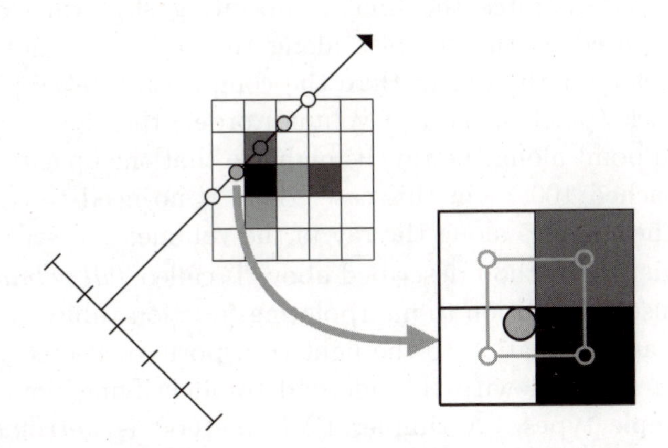

Figure 15.9. Schematic of interpolation in a cell created from voxels neighboring the sample.

lation (Section 11.7). Figure 15.9 illustrates that the interpolation takes place over a cell defined by the voxels neighboring the sample. Many other, higher-order, interpolation solutions are possible as well. The scalar value obtained from interpolation defines the material at this sample. By using a transfer function, we can map scalar values to color (RGB) and opacity (A, for alpha blending), which are called *optical properties*. More detail on transfer functions is presented in Section 15.5.

Diagram 3 illustrates that lighting takes place next for each sample on the ray. As is discussed in Section 16.3.1, a gradient is needed defining the orientation at this volume point in order to compute the lighting with respect to the light sources. Gradients may be computed several ways. One simple method would be to compute gradients at the voxel centers, perhaps using a divided difference method (as discussed in Section 9.4). Then the gradients can be interpolated in the same fashion as the function values. Another part of the lighting equation involves the material properties, which were obtained from the transfer function in the previous step. As discussed in Section 16.3, lighting means computing one or more of the following: reflection, refraction, emission, transmission, and absorption. By adding this extra step of illumination, the volume rendering will have better depth perception and greater surface structure contrast.

Diagram 4 illustrates the final compositing step whereby the lighting computed at the samples along the ray are combined to define one color for the pixel. Here the compositing takes place in a front-to-back (pixel to back of volume) order; therefore, at some intermediate point along the ray, it might be that the opacity of the pixel has reached 100%—in this case, there is no need to continue processing the samples along the ray in the volume.

The rendering method described above is called *full volume rendering* because, in addition to interpolating function values, we have included an approximation to the light transport equations. Isosurfaces can be achieved with this method by identifying boundaries between sample types. A simpler DVR method is *maximum intensity projection* (MIP) in which only the maximum value of all samples is written to the pixel. No emission or absorption values are calculated; rather, the sample values are scaled to a gray scale $[0, 1]$. This method has been a favorite for small feature identification, such as in CT angiography and evaluating vasculature; however, a disadvantage is poor depth information. Between full volume and MIP is *x-ray rendering*; for this method, all sample values are summed. This is typical (today) for diagnostic medical imaging, and the final result looks like a traditional x-ray. MIP and x-ray are faster to compute than full volume rendering, but lack its robustness. For instance, MIP images make the task of identifying relative depths difficult.

The methods described above are called *image-order techniques* because the computation is driven by the image (pixels) rather than the objects in the data set. But there are also *object-order techniques.* One such technique is called *splatting.* This method works by decomposing the volume into basis functions (see Section 8.6), and then using a weighting scheme—the result is that the functions are pushed (i.e., "splatted") onto the image plane.

15.5 Transfer Functions

The transfer function pairs materials, specified by scalar values, with color (emission) and opacity (absorption). For DVR, this pairing is key to identifying important features and differentiating materials through the assignment of correct optical properties. This process of

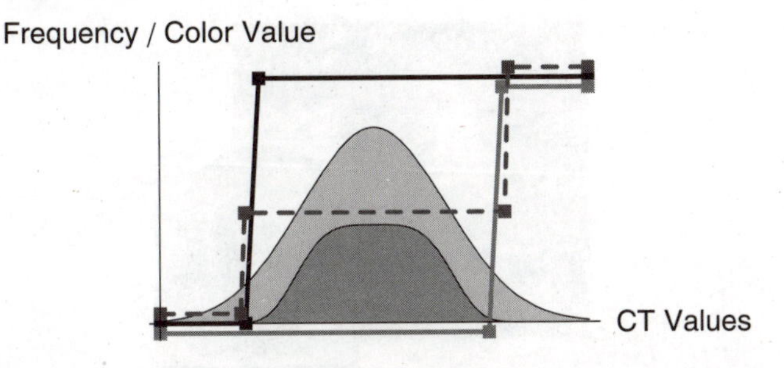

Figure 15.10. The transfer function to isolate blood in a CT scan is superimposed over the scan's histogram, the gray area. Values that correlate to blood are a darker shade of gray. The black polygon corresponds to red; the transfer function will return red for values in the "blood" range. Blue and green share the same graph with the light gray polygon. Opacity, denoted by the dashed line, is set as a piecewise constant over the scan. Bone values are to the right; therefore, the transfer function returns 100% opaque there.

mapping data acquisition readings, such as radiodensities, to a transfer function is called *material classification*. Figure 15.10 presents an example of a transfer function that is intended to highlight blood in a CT scan.[6] As the figure shows, the transfer function is actually made up of four functions: one each for red, green, blue, and opacity (transparency). Generally, the practice is to design the transfer function in conjunction with the histogram that records frequency of values in the scan. Some specialists have the experience to isolate materials by looking at these histograms.

Figure 15.11 illustrates that the transfer function can be used to display some materials but not others. An understanding of the properties of the materials in the volume is important. All materials present need to be assigned color and opacity ranges. Bone, for instance is white and rather dense so it is made 100% opaque. Tissue is typically colored pink to red, and it might take on a range of opacities, say 80–90%. If we want to ignore a material (and not render it), it can be given opacity 0%, and it will be invisible. Last, the space outside the volume should be given a color. Also considered is the position of the material in the volume. For example, is the material on the outside layer? This is called *material occupancy*.

[6]The histogram in this example is a simplified version of a real scan.

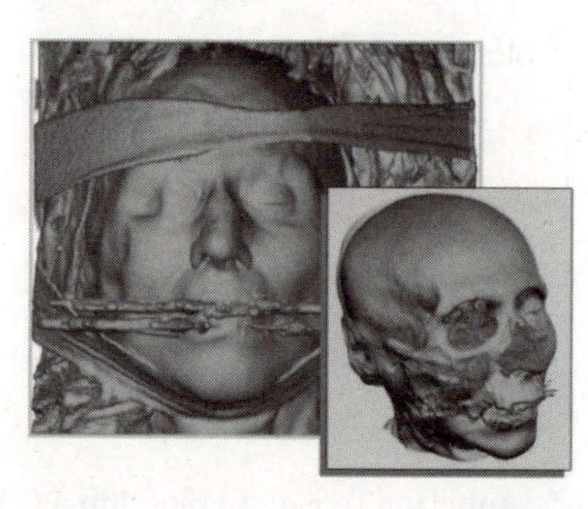

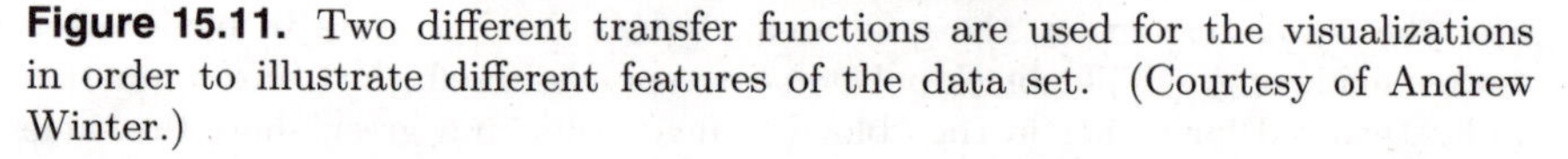

Figure 15.11. Two different transfer functions are used for the visualizations in order to illustrate different features of the data set. (Courtesy of Andrew Winter.)

The transfer functions is typically fine-tuned to highlight the phenomena of interest. If the transfer functions are not designed properly, however, an important but unknown feature in the data might be missed. Transfer function design is currently a field of research in its own right. Computer scientists, working with domain scientists, research the best sets of functions for various phenomena. Often, transfer functions are created by examining sample slices within a software tool specially designed to create transfer functions—a trial-and-error process. Much research is being done, however, on finding methods for semiautomatic and automatic transfer function definition. The example in Figure 15.10 is a *1D transfer function*. Research is taking place on *multidimensional transfer functions* that can better handle values that represent variable amounts of overlapping or mixed materials.

15.6 Comparison of Contouring and DVR

Figure 15.12 illustrates the difference between volume visualization with contours and DVR. The perceived sharpness in the contoured image is misleading because the fine structures, revealed in the DVR image, are missing. Marching cubes creates structures that are hanging in space; those are clearly not desirable. Small features and branches are generally difficult to detect properly by using marching cubes. Also, amorphous phenomena cannot be adequately represented by isosurfaces.

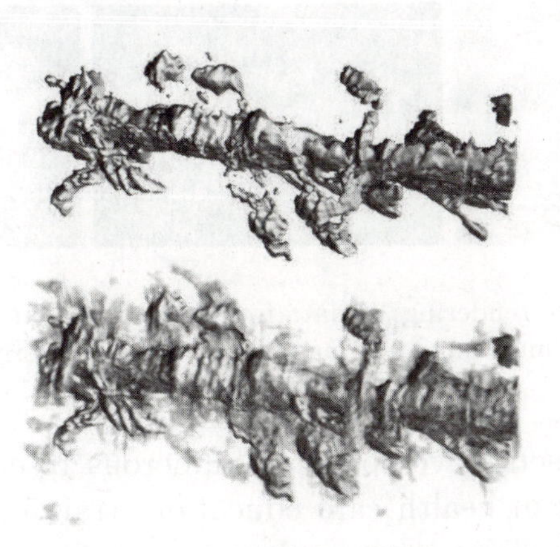

Figure 15.12. A comparison of contouring (top) and DVR (bottom). (Courtesy of Andrew Winter.)

However, contouring has the advantage of computing geometry that can be rendered with the standard graphics pipeline, and thus rotated for observation from any viewpoint. Additionally, contouring typically is faster than DVR; particularly if we take into account that there is no need to reprocess for a new image. On the downside, a large number triangles might be required to represent a data set.

DVR viewing is typically orthographic because perspective is more difficult to compute and can introduce problems. (Orthographic tends to be better for many applications due to its consistency in lengths.)

15.7 Case Study: Visible Human Project

The Visible Human Project is a National Institutes of Health (NIH) initiative to create a complete, anatomically detailed, 3D representation of a normal male and female human body. Data sets have been acquired for a male, sectioned at one millimeter intervals, and a female, sectioned at one-third millimeter intervals, by using CT, MRI, and cryosection images.[7] The female data set, for instance, is about 40 GB in size. Figure 15.13 illustrates three renderings of data from the Visible Human Project.

[7]Cryosection means that the body was frozen, sliced, and then imaged!

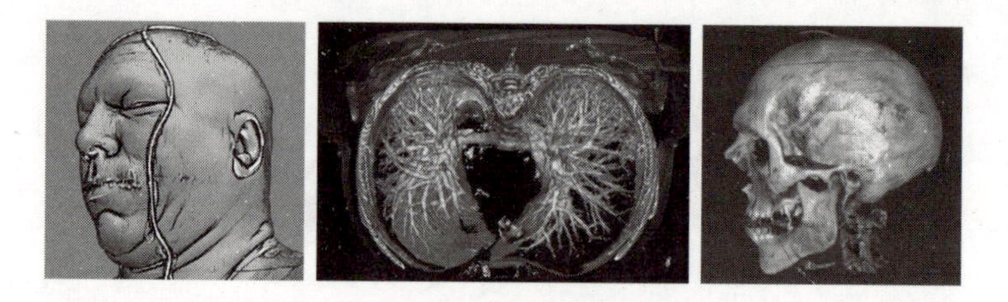

Figure 15.13. Three renderings of data from the Visible Human Project. (Courtesy of Scientific Computing and Imaging Institute, University of Utah)

These data sets have spawned numerous research projects, including projects in health care education, training applications for health care and disaster management, virtual reality and surgical simulation, and testbeds for biomedical research. Links to these projects and others may be found on the project's website, http://www.nlm.nih.gov/research/visible/visible_human.html. Nontrivial data sets such as these provide benchmarks for researchers developing algorithms.

15.8 Data Cutting

Data cutting refers to the process of cutting through a volume with a surface and displaying the function values defined on the surface. In Figure 15.14, the cutting surface is a plane; this is typical but not necessary. A reasonably effective method for simulating full vol-

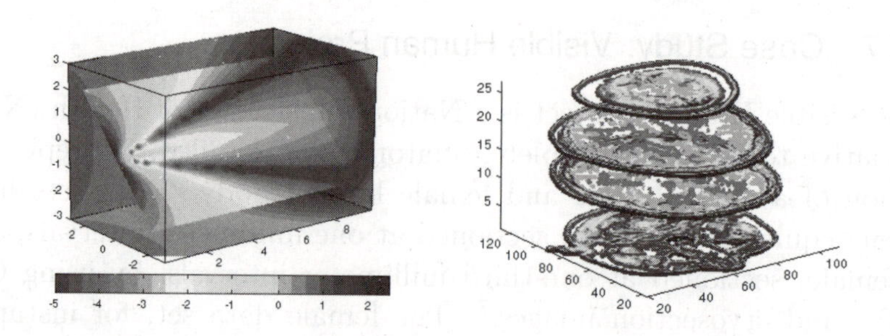

Figure 15.14. Examples of planar cutting surfaces: three planes on the faces of the volume with one plane at an arbitrary angle (left) and multiple planes, stacked (right). (Courtesy of MATLAB from MathWorks.)

ume rendering is to use many cutting surfaces displayed with some amount of transparency, and then to render the surfaces back to front. Data cutting is a form of *probing* or resampling the data set to find regions of interest.

The left example of Figure 15.14 displays planar faces of a volume. This gives a sort of frame of reference for the behavior of the function in the volume. In Section 15.9, we look at methods to visualize vector data; these methods marry well with such cutting visualizations.

15.9 Vector Fields

In Section 11.1, we looked at computing a special type of vector field over bivariate functions, called a gradient field. We visualize these vector fields with small arrows drawn at discrete grid points, sometimes called *arrow plots.* A vector is defined by a direction and magnitude. Communicating both pieces of information effectively typically involves variation in vector length, color, or line width. This method can be extended to trivariate functions as well, but the visualizations can quickly become difficult to interpret.

A *hedgehog plot*, as illustrated in Figure 15.15, is a special case of a 3D vector field plot. In this figure, triangle (unit) normals are

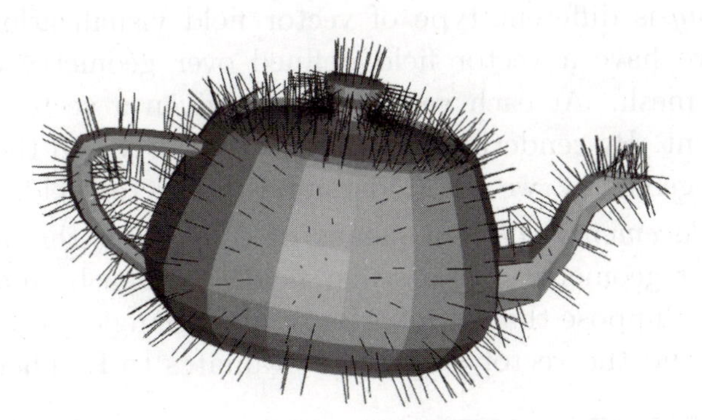

Figure 15.15. Hedgehog plot: triangle normals are drawn at the centroid of each triangle.

drawn at the centroid of each triangle in a triangle mesh.[8] The normals are drawn simply as straight lines, but an arrow could have been used instead. Also noteworthy is the fact that the normals have been scaled to somehow match the dimension of the object. If they were too short, they would not be visible; if they were too long, the plot would be a dark mess. These sticks, or arrows, are an example of *oriented glyphs*, which are symbolic figures or shapes that encode more than one dimension. The type and dimension of data to be visualized determines the appropriate type of glyph—the possibilities are endless. Glyphs are commonly used in information visualization; more information may be found on the dangers of using them in the literature of that field. For instance, inappropriate scaling of a glyph can unintentionally miscommunicate.

The methods above could be called point-based visualization because the vector field is considered at a point and translated to a glyph. Another class of methods is *particle tracing*, which works with characteristic curves. One type of particle tracing method is called streamlines, which is a method of tracing trajectories in a vector field. Streamlines were generated from 3D vector data, for example, in the tornado visualizations in Figure 15.16.

Many variations of streamlines have been developed, and each is designed to highlight a particular feature in a flow. For instance, streamribbons show particle trajectory and rotation through the use of a ribbon that represents the streamline. The list goes on: streamtubes, streamballs, stream surfaces, and so on!

Warping is different type of vector field visualization method. Suppose we have a vector field defined over geometry defined by a triangle mesh. At each vertex, we are given a vector defining a displacement. By rendering the original geometry and the displaced or warped geometry, we can understand the vector field.

A *displacement plot* is a means of converting the vector field defined over geometry into a scalar field that can be mapped to a color map. Suppose that at a vertex $\mathbf{v}$ of a triangle mesh, a normal $\mathbf{n}$ is given and the vector field there evaluates to $\mathbf{f}$. Then $s = \mathbf{n} \cdot \mathbf{f}$

[8] Hedgehog plots, or normal plots, are useful for checking whether the orientation of triangles in the triangle mesh are consistently defined; this is a recommended exercise if the shading appears incorrect.

Figure 15.16. Tornado visualization demonstrating 3D vector fields. (Courtesy of Scientific Computing and Imaging Institute, University of Utah.)

can be used to determine whether the motion of the vector field at **v** is in the direction of the surface normal.

Often, the purpose of vector field visualization methods is to see the general behavior of the flow. Sometimes, additional scalar values will need to be visualized along with the vector field. For example, temperature might accompany a vector field defining air flow in which the vectors define flow direction and speed.

Vector field analysis and visualization is a very important and challenging area. The visualization community is still developing the tools scientists and engineers need to gain insight into the massive amounts of experimental measurements and numerical simulations. We cannot attempt to cover this topic here. The examples above should open the door, though, to imagining the possibilities.

15.10 Tensor Fields

Informally, a tensor field, or *tensor*, is a multidimensional array describing some quantity that varies from point to point. In visualization, the current focus is on 3×3 tensors.[9] This type of tensor is important in physics and engineering. Two commonly analyzed tensors are the stress and strain tensors. There is much more to know about tensor fields, but that is beyond the scope of this text. Let's

[9]These are called rank 2 tensors. Rank 0 and rank 1 tensors correspond to scalar and vector fields, respectively.

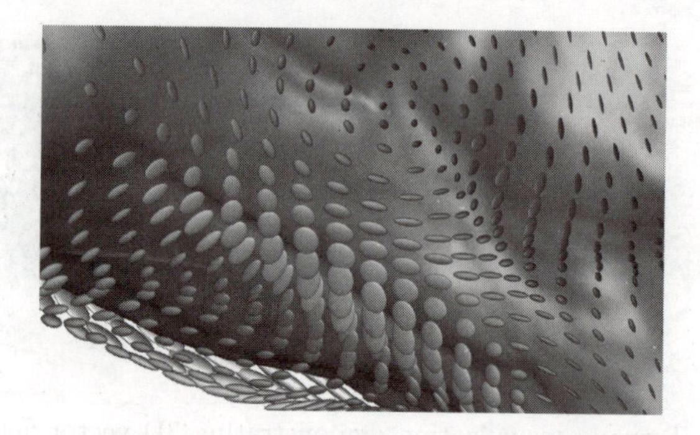

Figure 15.17. Tensor ellipsoids. (Courtesy of Scientific Computing and Imaging Institute, University of Utah.)

look at the stress tensor, however, to get a glimpse of the visualization challenges.

The *stress tensor* measures force per unit area. The distribution of force is broken down into its shear and normal components. The normal component, σ_{ii}, acts perpendicularly to the surface, and the shear component, τ_{ij}, acts tangentially to the surface. Together they form the stress tensor,

$$S = \begin{bmatrix} \sigma_{xx} & \tau_{xy} & \tau_{xz} \\ \tau_{yx} & \sigma_{yy} & \tau_{yz} \\ \tau_{zx} & \tau_{zy} & \sigma_{zz} \end{bmatrix}.$$

How do we visualize a matrix at points on a surface or in a volume? A first solution is to simplify the problem. An eigenvector and eigenvalue analysis of S can be helpful when they are of physical significance, which is the case for stress tensors. (See Section 6.1 for a discussion on eigenvectors.) Recall that the eigenvectors are mutually perpendicular. This opens the door to visualization of *tensor ellipsoids* (see Figure 15.17). An ellipsoid's minor, medium, and major axes are in the directions of the eigenvectors, and the lengths are determined by the eigenvalues. These are a form of a glyph.

15.11 Haptic Visualization

The word "haptic" refers to the sense of touch. The field of haptic technology, or haptics, refers to the technology that allows a user to

Figure 15.18. Virtual colonoscopy: haptic visualization of polyps in the walls of the intestine. Virtual interactions are done with a Phantom Desktop device attached to a laproscopic device. (Courtesy of A. Sridaran and K. Kahol, CUbiC, Arizona State University.)

touch virtual (haptic) objects. These objects are defined by their geometry and forces. A haptic device is held by a user, and with this device the user can explore an object through its forces. The user may not necessarily be able to see the object on the screen. As the haptic device touches the virtual object, tactile or force feedback is returned through the device. This might be a vibration or resistance. The computation of these forces is called haptic rendering. Haptics has applications in many fields, some of which include visualization for the blind, enhanced gaming experiences, or as an aid to medical training via virtual surgeries. Figure 15.18 illustrates the latter. Haptic visualization allows for an additional modality for exploring data sets. It is known that sight allows us to most quickly absorb information; however, haptics can enhance a visualization or experience. Coupled with visualization, haptics can increase realism, and for some applications it can improve operator performance. The technology is still a limiting factor, but as advances are made, new applications will arise.

15.12 Problems and Experiments

1. Using a software product such as Mathematica, experiment with contours $f(x, y, z) = c$ for the following functions.

$$f(x, y, z) = \sin(x) + \sin(y) + \sin(z) \quad \text{and}$$
$$f(x, y, z) = x^2 - y^2 + z^2$$

Try to guess what the contours will look like before plotting.

2. Figure 15.9 illustrates bilinear interpolation. Write down an expression for the point of interpolation given gray values at the vertices of the cell. Experiment with a method that uses more information.

3. See equation (16.7) for a compositing equation. Experiment with the compositing step in direct volume rendering by providing colors and alpha-values. (See Section 16.1 for more information on color models.)

4. Figure 15.9 illustrates bilinear interpolation for determining the color at a point. Experiment with this method and compare it to biquadratic interpolation.

5. Experiment with automatically determining a reasonable length for the vectors in a hedgehog plot.

6. Create a figure illustrating warping and displacement plots.

16

Background: Computer Graphics

This chapter is designed to be a tutorial on computer graphics techniques, which are the core building blocks for scientific visualization.

First we give an introduction to color models because they are key to a robust visualization. Next is a section describing key elements of the graphics pipeline. This section describes the transformations needed to convert a 3D triangle mesh to 2D pixels on the screen, resulting in the desired orientation and giving a 3D appearance. To give the mesh more realism, we apply an illumination model. Three techniques are described: local illumination, which can be computed in real time; global illumination, which generally is not computed in real time but results in more realistic rendering than local methods; and nonphotorealistic rendering (illumination), which simulates methods such as cartoon or hand-drawn renderings. Texture mapping is introduced next as a method to create complexity in an image without the expense of geometric complexity. Underlying the entire visualization process are the methods of sampling, smoothing, and reduction. These concepts are introduced in the last section.

More information on the material in this chapter is available in the OpenGL Programming Guide [17] or Shirley et al. [16].

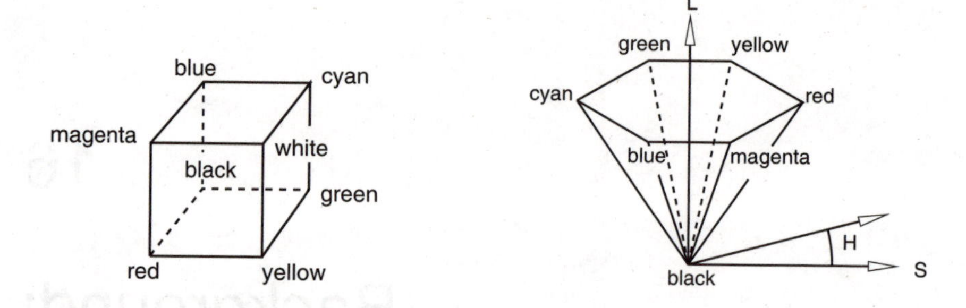

Figure 16.1. The RGB color model has primary colors red, green, and blue represented by the three axes of the unit cube; gray scales are on the line from black to white (left). The HSL color model represents hue in terms of an angle, saturation is a radius, and lightness moves up through the hexcone (right).

16.1 Color Models

Entire texts are written on the theory of color. Here we present only the basic concepts likely to be encountered when working in scientific visualization.

We perceive color through three cones in each receptor cell of our retinas. To each cone belongs a function that defines the amount of absorbtion of each wavelength of light. This *trichromatic theory* has motivated the design of several color models that represent visible light through three components. A key development turned out to be the use of the *RGB color model* by color TV, in which three phosphors: red, green, and blue, are combined to create colors. The RGB model carried over to computer screens, and it is the predominant model in computer graphics and scientific visualization. Figure 16.1 (left) illustrates that a color in this model can be thought of as a point in a unit cube with R, G, B axes. The *RGBA model* is the RGB model with the addition of transparency, or alpha value. See Section 15.5 for the use of transparency (opacity) in volume rendering. Examples of some colors expressed by R, G, B triples are given in Table 16.1.

In some software, the RGB components are specified as above, where each is a floating-point number in $[0, 1]$. In other software, each component is an integer in $[0, 255]$. In the hardware, each "color gun" is allocated eight bits, and thus has 2^8, or 256, intensity settings

red:	1, 0, 0	yellow:	1, 1, 0
green:	0, 1, 0	dark gray:	0.2, 0.2, 0.2
blue:	0, 0, 1	light gray:	0.8, 0.8, 0.8
white:	1, 1, 1	black:	0, 0, 0

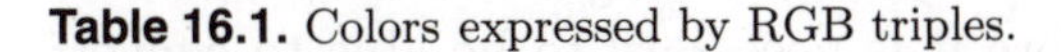

Table 16.1. Colors expressed by RGB triples.

for illuminating the phosphor. Therefore, when you specify color as a floating-point number in $[0, 1]$, it is actually mapped to an integer in $[0, 255]$. This really is not a limitation, though; 24-bit color, as it is called, can represent more than 16 million colors $(256 \times 256 \times 256)$. The number of bits used is called *color depth.* A 32-bit color system is normally referring to 24-bit color with an alpha channel.

Except for the most basic colors, it is difficult to specify color with the RGB model. Varying between shades of color takes some attention. The *HSL model* (hue, saturation, luminance) is better suited for intuitive color specifications. This model, illustrated in Figure 16.1 (right), lives in a cone. Colors such as red, green, or yellow are of different *hue*, H, and are placed around the circle; therefore, we assign $H \in [0, 360°]$. The dilution of a hue by white, such as the difference between blue and sky blue, is its *saturation*, $S \in [0, 1]$. And a color's achromatic brightness or intensity is called *lightness*, $L \in [0, 1]$. This model is well suited for a color-selection user interface.

Mirroring the RGB examples, a sample of HSL colors is given in Table 16.2.

Since graphics hardware requires RGB, a formula converting HSL to RGB must be used [3]. Interestingly, the top of the hexcone

red:	0°, 1, 1	yellow:	60°, 1, 1
green:	120°, 1, 1	dark gray:	0°, 0, 0.2
blue:	240°, 1, 1	light gray:	0°, 0, 0.8
white:	0°, 0, 0	black:	0°, 0, 0

Table 16.2. Colors expressed by HSL triples.

corresponds to the hexagon seen from a vantage point on the white-to-black diagonal of the RGB model. Subcubes of the RGB model are cubes with one corner at black and the opposing corner at gray $[c, c, c]$. Views of the subcubes from the same vantage point results in hexagons of varying lightness in the HSL model.

For printing and volume rendering, it might not be possible to work with color, and therefore it is necessary to be able to convert color to *gray scales*. This means that color $\mathbf{c} = [R, G, B]$ is converted to gray, $\mathbf{y} = [Y, Y, Y]$. Notice that $\mathbf{y}$ is a point on the "gray" line, $\mathbf{l}$, from black $[0, 0, 0]$ to white $[1, 1, 1]$. A simple but reasonable method for assigning a gray value is to let $Y = (R+G+B)/3$. The resulting $\mathbf{y}$ is the closest point on $\mathbf{l}$ to $\mathbf{c}$. A method that weights the RGB values based on our eye's response to constant luminance as wavelength (color) is varied sets

$$Y = 0.3R + 0.59G + 0.11B.$$

Our eyes are most sensitive to yellow-green light and least sensitive to blue-violet light. The value Y is part of the so-called YIQ model which is used for efficient transmission of RGB, and it is downward compatible with black and white TV.

See an Information Visualization text such as Ware [19] for studies on the best color combinations for readability and to best communicate visualizations to color-blind people.

The CMY (cyan, magenta, yellow) color model is important for printing. The primary colors in the CMY model are called subtractive because colors are specified by defining what is subtracted from white light. Cyan, magenta, and yellow are the complements of red, green, and blue, respectively. For example, when cyan ink is added to paper, no red light is reflected. Therefore, cyan is green mixed with blue.

16.2 The Graphics Pipeline

In this section, let's look at the steps required to display 3D geometry on a screen made up of a rectangular array of pixels. Each pixel is assigned one color. So how are these colors determined? This section only partially answers this question by describing the mapping of 3D geometry to 2D pixels. Section 16.3 shows how the color is determined.

For the trip down the pipeline, all geometry must be converted to points, lines, or triangles. These are called *primitives*. Thus, no matter how smooth an object looks on the screen, it consists of primitives only!

The graphics pipeline involves several coordinate systems, namely

$$\text{object} \rightarrow \text{world} \rightarrow \text{eye} \rightarrow \text{clip} \rightarrow \text{NDC} \rightarrow \text{window},$$

where NDC means normalized device coordinates (see Section 16.2.2). The arrows indicate transition from one system to the next. These coordinate transformations have been designed to optimize the algorithms applied to a primitive traveling down the pipeline. A definition of each coordinate system will be introduced in the sections to follow.

Suppose we create one atom structure in a modeling package such as Maya. The atom would be said to live in *object coordinates*. This is typically a coordinate system that makes defining the geometry easy. In most cases, the object will be centered about the origin. Now suppose we want to orient many atoms in space to form a particular pattern. The coordinates of the atoms are now in *world coordinates*. Simple 2D or 3D affine transformations, such as translation, rotation, or scaling, might be used to transform the coordinates from object to world coordinates.

In Section 16.2.1, the discussion focuses on the transformation from world to eye coordinates. This so-called viewing transformation is commonly described in terms of setting up a camera. Section 16.2.2 focuses on the transformations from eye to clip coordinates and clip to NDC coordinates.[1] These two transformations prepare the vertices for the final projection into the window.[2] The entire drawable area for a graphics application is the window. A rectangular subwindow is called a *viewport*; sometimes these terms are interchanged.

---

[1]We say "NDC coordinates," which does repeat a word, but that's just they way it's done!

[2]The stages from eye to NDC are called projection, but a projection as we normally think of it—a flattening—doesn't really occur until later in the pipeline. We look at why this is so in Section 16.2.5.

If we use a graphics application, such as Mathematica, we won't need to know all the details of the graphics pipeline. However, Sections 16.2.1 and 16.2.2 could be considered a "case study" for linear algebra!

16.2.1 Setting up the Camera

Let's assume that we have placed the geometry that we would like to display in a world coordinate system. Soon we will want to define the type of projection (orthographic or perspective) and the specifics of the projection. Standard practice is to make the xy-plane the projection plane. Thus to facilitate defining the projection step, we move our geometry to sit "centered" on the $-z$-axis, and these new coordinates are called *eye coordinates.* Just how we move and orient our geometry into eye coordinates is conveniently defined by parameters analogous to setting up a camera in the world coordinate system.

Specifically, setting up a camera involves choosing its

- location, $\mathbf{e}$ (called the *eye point*),
- line of sight, defined by the line through the eye and a point $\mathbf{a}$ (called the *at point*), and
- orientation, $\mathbf{u}$ (called the *up vector*).

These parameters are illustrated in Figure 16.2 (left). There are a few restrictions on these camera parameters. The eye and at points must not be identical. The up vector determines how the camera is rotated. For instance, will the final image appear horizontally or vertically? The up vector does not need to be orthogonal to the line of sight, $\mathbf{a} - \mathbf{e}$. Obviously, the up vector also must not be the zero vector.

Now we have the means for describing the orientation of our geometry on the $-z$-axis. The viewing operation will take

- the eye point to the origin,
- the at point onto the $-z$-axis, and
- the up vector will lie in the $+yz$-plane.

This is illustrated in Figure 16.2 (right).

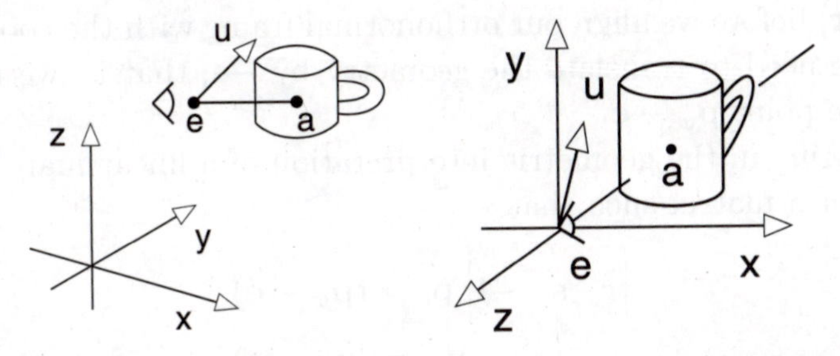

Figure 16.2. Setting up a camera in world coordinates (left) and the camera's position in eye coordinates (right).

Viewing Transformation Matrix. Suppose a vertex of our geometry in world coordinates is labeled $\mathbf{p}_w$, and the corresponding point in eye coordinates is labeled $\mathbf{p}_e$. Using our knowledge of the connection between linear maps and coordinate transformations, the world to eye transformation is very easy to construct!

To review: recall that a matrix $A = \begin{bmatrix} \mathbf{a}_1 & \mathbf{a}_2 & \mathbf{a}_3 \end{bmatrix}$ defines a linear map that maps a vector $\mathbf{v}$ in the $[\mathbf{e}_1, \mathbf{e}_2, \mathbf{e}_3]$-coordinate system to a vector $\mathbf{v}'$ in the $[\mathbf{a}_1, \mathbf{a}_2, \mathbf{a}_3]$-coordinate system,

$$\mathbf{v}' = A\mathbf{v}.$$

Our problem here is simply a coordinate transformation, so the goal is to stage it in that context.

The first step is to form an orthonormal frame (a set of orthogonal unit vectors) from the camera parameters:

$$\mathbf{l} = \frac{\mathbf{a} - \mathbf{e}}{\|\mathbf{a} - \mathbf{e}\|}, \qquad \mathbf{r} = \frac{\mathbf{l} \wedge \mathbf{u}}{\|\mathbf{l} \wedge \mathbf{u}\|}, \qquad \mathbf{s} = \mathbf{r} \wedge \mathbf{l},$$

where $\wedge$ is the cross product. The vector $\mathbf{l}$ is called the *line of sight vector.* Notice that $\mathbf{r}$ is on our right as we look down the line of sight, and then $\mathbf{s}$ will correspond to the up direction. The vectors $\mathbf{u}$ and $\mathbf{l}$ form a plane. In this plane, the side on which $\mathbf{u}$ lies determines the resulting orientation: right-side-up or up-side-down. The angle between $\mathbf{u}$ and $\mathbf{l}$ is inconsequential.

The mapping of the orthonormal frame that we want is

$$\mathbf{l} \rightarrow -\mathbf{e}_3 \qquad \mathbf{r} \rightarrow \mathbf{e}_1 \qquad \mathbf{s} \rightarrow \mathbf{e}_2.$$

However, before we align our orthonormal frame with the coordinate axes, we need to translate the geometry by $-\mathbf{e}$, that is, we need to form the point $\mathbf{p}_w - \mathbf{e}$.

Bringing in the geometric interpretation of a linear map, we can construct a matrix such that

$$\begin{bmatrix}\mathbf{r} & \mathbf{s} & -\mathbf{l}\end{bmatrix}\mathbf{p}_e = (\mathbf{p}_w - \mathbf{e})$$
$$A\mathbf{p}_e = (\mathbf{p}_w - \mathbf{e}).$$

However, $\mathbf{p}_e$ is unknown, so we need to find the matrix A^{-1} that forms

$$\mathbf{p}_e = A^{-1}(\mathbf{p}_w - \mathbf{e}).$$

Now we can take advantage of having formed an orthonormal frame because $A^{-1} = A^{\mathrm{T}}$. Hence,

$$\mathbf{p}_e = A^{\mathrm{T}}(\mathbf{p}_w - \mathbf{e}) = \begin{bmatrix}\mathbf{r}^{\mathrm{T}}\\ \mathbf{s}^{\mathrm{T}}\\ -\mathbf{l}^{\mathrm{T}}\end{bmatrix}(\mathbf{p}_w - \mathbf{e}), \qquad (16.1)$$

and we have our transformation to eye coordinates.

A Numerical Example. We now look at a simple example to keep the calculations easy to follow. Suppose our modeling transformations place our object on the $-x$-axis, as illustrated in Figure 16.3 (left). Let's set up our camera with the following parameters:

$$\mathbf{e} = \begin{bmatrix}0\\0\\0\end{bmatrix}, \qquad \mathbf{a} = \begin{bmatrix}-1\\0\\0\end{bmatrix}, \qquad \mathbf{u} = \begin{bmatrix}0\\0\\1\end{bmatrix}.$$

From this camera set-up, we form the following orthonormal frame

$$\mathbf{l} = \begin{bmatrix}-1\\0\\0\end{bmatrix}, \qquad \mathbf{r} = \begin{bmatrix}0\\1\\0\end{bmatrix}, \qquad \mathbf{s} = \begin{bmatrix}0\\0\\1\end{bmatrix},$$

which is also illustrated in Figure 16.3 (middle).

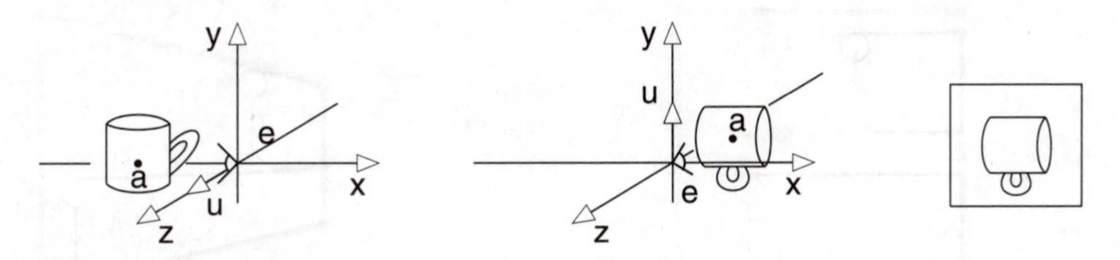

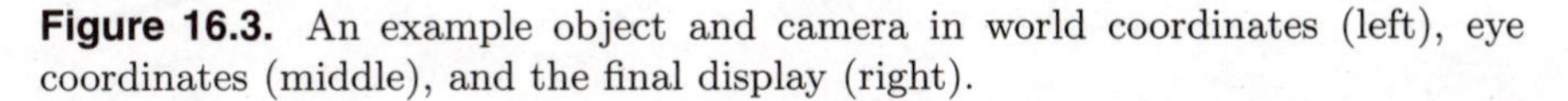

Figure 16.3. An example object and camera in world coordinates (left), eye coordinates (middle), and the final display (right).

Since the camera is already located at the origin, a translation is not necessary and

$$A = \begin{bmatrix} 0 & 0 & 1 \\ 1 & 0 & 0 \\ 0 & 1 & 0 \end{bmatrix}.$$

Thus the eye coordinate points $\mathbf{p}_e$ are calculated as

$$\mathbf{p}_e = \begin{bmatrix} 0 & 1 & 0 \\ 0 & 0 & 1 \\ 1 & 0 & 0 \end{bmatrix} \mathbf{p}_w.$$

Let's check that our original goals—that the at point is mapped to the $-z$-axis and the up vector is mapped to the $+yz$–plane—are satisfied.

$$A^{\mathrm{T}}\mathbf{a} = \begin{bmatrix} 0 \\ 0 \\ -1 \end{bmatrix} \quad \text{and} \quad A^{\mathrm{T}}\mathbf{u} = \begin{bmatrix} 0 \\ 1 \\ 0 \end{bmatrix},$$

so they do indeed.

The final view of the object is illustrated in Figure 16.3 (right).

16.2.2 Projections in the Graphics Pipeline

Now our object lives in eye coordinates: we are positioned at the origin looking down at the object positioned on or near the $-z$-axis.

Orthographic and perspective projections are the methods most commonly used in computer graphics. Figure 16.4 illustrates both

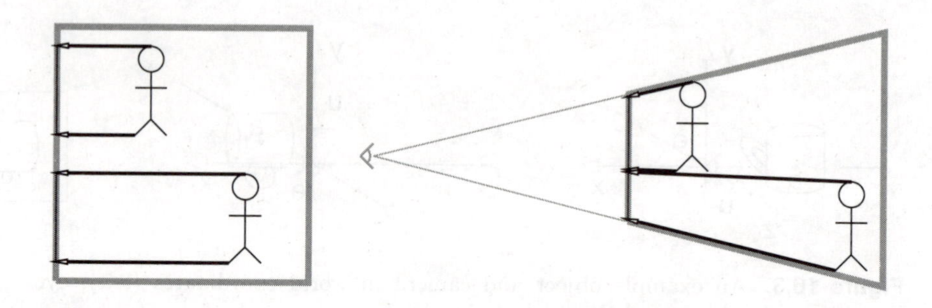

Figure 16.4. An orthographic projection in which vertices are projected perpendicular to the view plane (left), and the same view with a perspective projection (right), in which vertices are projected into the view plane through the center of projection, resulting in foreshortening (right).

methods. At the end of the pipeline, the vertices will be projected into a *view plane*, which in turn is mapped to the window.

An orthographic projection projects all vertices in a direction perpendicular to the view plane. Architects and engineers tend to prefer orthographic images because distances and angles are preserved. However, these images can make it difficult to "see" an object as 3D, and more than one view is commonly needed.

A perspective projection projects a vertex along a line defined by that vertex and the eye, or *center of projection (cop)*. This results in a foreshortening effect: objects farther away appear smaller than objects the same size that are closer to the eye. This foreshortening makes perspective very good for creating realistic images. Orthographic projection can be thought of as a special case of perspective: the center of projection is located infinitely far away.

In the graphics pipeline, specifying a projection answers two questions:

1. How is the geometry projected?

2. What is displayed?

Part of specifying a projection includes defining the parameters of a *viewing volume*. In Figure 16.4, observe that the orthographic viewing volume is a (rectangular) box and the perspective viewing volume is a *frustum*, or truncated pyramid. When we have our clip coordinates, all vertices outside of the viewing volume will be clipped, or eliminated from the list of vertices continuing down the pipeline.

The transformation from eye to clip coordinates is controlled by a matrix that is often called the projection matrix. The key idea is that when we arrive in clip coordinates, regardless of whether we want an orthographic or perspective projection, all our vertices live in a *normalized 4D cube.* (We discuss how this works in the sections to follow.) The transformation from clip to NDC coordinates is the "perspective division" step (see Section 16.2.3) which maps the 4D cube into the 3D cube with lower-left and upper-right vertices

$$\begin{bmatrix} -1 \\ -1 \\ 1 \end{bmatrix} \quad \text{and} \quad \begin{bmatrix} 1 \\ 1 \\ -1 \end{bmatrix},$$

respectively.[3] This process of bringing the geometry into this special volume is called *normalization.* This special volume is often called the *canonoical viewing volume.*

A parallel projection into the $z = 0$ plane is all that is needed to create the 2D vertices for rasterization, which is the conversion of 2D primitives into pixels. (This amounts to ignoring the z-value.) Importantly though, the z-values must be available at this stage in the pipeline so we know what geometry is closest to the eye and thus visible. This is called z-buffer hidden surface removal (more on this in Section 16.2.5).

Let a homogeneous point in eye coordinates be $\bar{\mathbf{p}}_e$, its corresponding point in clip coordinates be $\bar{\mathbf{p}}_c$, and its corresponding (affine) point in NDC coordinates be $\mathbf{p}_{ndc}$. (See Section 3.5 for an introduction to homogeneous coordinates.)

Orthographic Projections. An orthographic projection is defined by the lower-left and upper-right vertices of the viewing volume, respectively:

$$\begin{bmatrix} l \\ b \\ -n \end{bmatrix} \quad \text{and} \quad \begin{bmatrix} r \\ t \\ -f \end{bmatrix}.$$

The orthographic projection matrix translates and scales this box (actually the vertices within) to its normalized 4D box with lower-left

[3]The "lower-left" and "upper-right" are relative to our eye looking down the $-z$-axis.

and upper-right vertices, respectively:

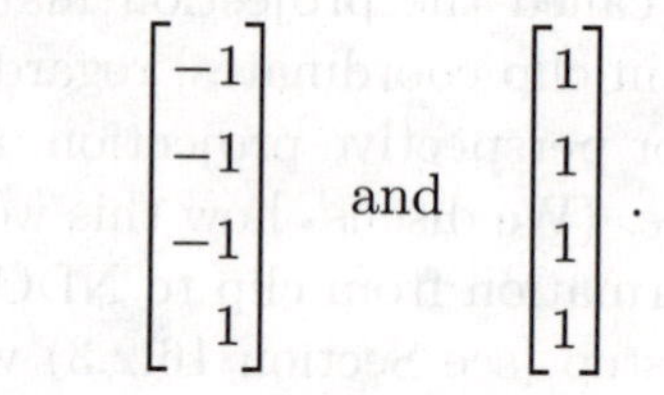

$$\begin{bmatrix} -1 \\ -1 \\ -1 \\ 1 \end{bmatrix} \quad \text{and} \quad \begin{bmatrix} 1 \\ 1 \\ 1 \\ 1 \end{bmatrix}.$$

The center of the box is

$$\mathbf{c} = \begin{bmatrix} (1/2)(l+r) \\ (1/2)(b+t) \\ (1/2)(n+f) \end{bmatrix}.$$

Thus, for an orthographic projection, the transformation from eye coordinates to clip coordinates is

$$\bar{\mathbf{p}}_C = \begin{bmatrix} 2/(r-l) & 0 & 0 & 0 \\ 0 & 2/(t-b) & 0 & 0 \\ 0 & 0 & 2/(n-f) & 0 \\ 0 & 0 & 0 & 1 \end{bmatrix} \begin{bmatrix} 1 & 0 & 0 & -c_x \\ 0 & 1 & 0 & -c_y \\ 0 & 0 & 1 & -c_z \\ 0 & 0 & 0 & 1 \end{bmatrix} \bar{\mathbf{p}}_e. \tag{16.2}$$

The product of the matrices in (16.2) is the *orthographic projection matrix*

$$M_o = \begin{bmatrix} 2/(r-l) & 0 & 0 & -(r+l)/(r-l) \\ 0 & 2/(t-b) & 0 & -(t+b)/(t-b) \\ 0 & 0 & 2/(n-f) & -(f+n)/(n-f) \\ 0 & 0 & 0 & 1 \end{bmatrix}.$$

Perspective Projections. The parameters used to describe the frustum are illustrated in Figure 16.5. It is common practice to draw the frustum in a 2D view in the yz-plane. The center of projection is at the origin. The angle θ is called the *field of view* and n is the distance from the eye to the near plane. This plane corresponds to the view plane. Then f is the distance to the far plane. The height

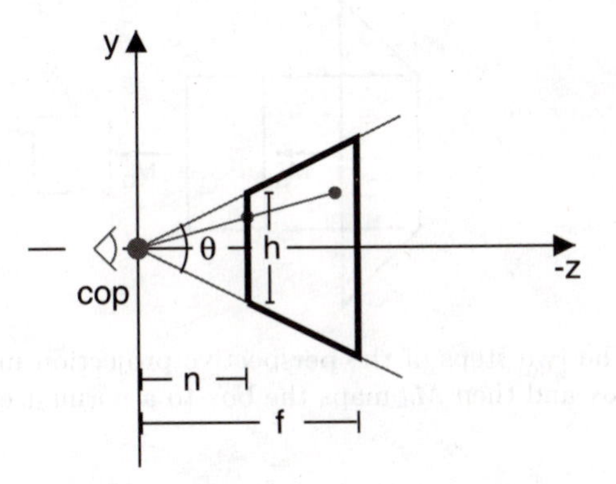

Figure 16.5. The perspective projection frustum, defined by the parameters θ, n, f, h and the width of the view (near) plane.

of the near plane, is denoted by h. These parameters are dependent, and they are related by

$$\tan(\theta/2) = \frac{h/2}{n}.$$

A human's field of view is roughly 65°. We can simulate a wide angle lens by creating a field of view that is large, for example, 80°. And a telephoto lens can be simulated with a small field of view, for example, 30°. This isn't an entirely correct analogy because the focal length of a camera is actually more closely related to the near distance. However, since our hypothetical camera has a variable film size, relating the field of view to the lens is simpler.

The most common and easiest method for defining the frustum is to define the parameters θ, a, n, f, where a is the aspect ratio (width/height) of the near plane. From these parameters we can calculate any of the others that define the frustum:

$$h = 2n\tan(\theta/2) \quad \text{and then} \quad w = ha.$$

Commonly, we choose the aspect ratio of the world coordinates to equal the aspect ratio of the viewport. This relationship assumes symmetry about the $-z$-axis.

The action of the *perspective projection matrix* is to map the frustum to its normalized 4D box with lower-left and upper-right

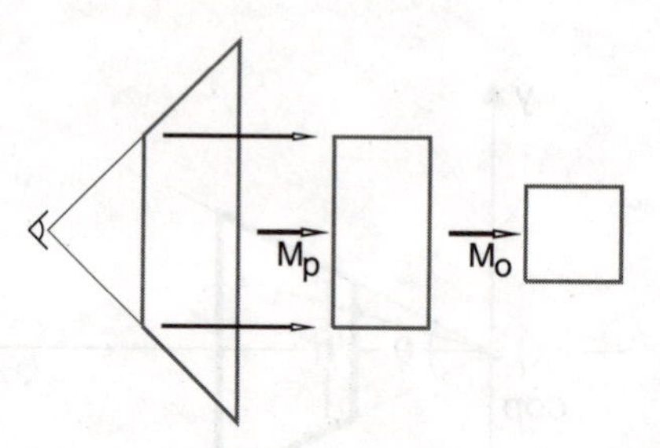

Figure 16.6. The two steps of the perspective projection matrix: M_p maps the frustum to a box and then M_o maps the box to a normalized box.

vertices

$$\begin{bmatrix} -n \\ -n \\ n \\ n \end{bmatrix} \quad \text{and} \quad \begin{bmatrix} f \\ f \\ -f \\ f \end{bmatrix}, \tag{16.3}$$

respectively. As illustrated in Figure 16.6, this takes place in two steps: M_p is a matrix that maps the frustum to a box with the same parameters as the frustum (l, r, b, t, n, f). Next, M_o is the orthographic mapping to take this general box to the normalized 4D box in (16.3).

The mapping M_p is a projective map, which preserves the following properties.

- *Lines map to lines.* If three points are collinear in eye coordinates, then they will be collinear in clip coordinates.
- *Planes map to planes.* If four points are coplanar in eye coordinates, then they will be coplanar in clip coordinates.
- *The inverse of the map exists.* This helps with screen picks—window coordinates can be transformed back through the pipeline to world coordinates.
- *Relative z-depth is preserved.* A point closer to your eye than another point in eye coordinates will also be closer in clip coordinates.

This perspective projection matrix takes the form:

$$M_p = \begin{bmatrix} 1 & 0 & 0 & 0 \\ 0 & 1 & 0 & 0 \\ 0 & 0 & (n+f)/n & -f \\ 0 & 0 & 1/n & 0 \end{bmatrix}.$$

Let's look at M_p's action on a point $\mathbf{x}$:

$$M_p\bar{\mathbf{x}} = \begin{bmatrix} x \\ y \\ z(\frac{n+f}{n}) - f \\ z/n \end{bmatrix}. \tag{16.4}$$

To understand the geometry, let's divide by the homogeneous "w" coordinate, and look at the corresponding affine point

$$\mathbf{q} = \begin{bmatrix} x \cdot n/z \\ y \cdot n/z \\ n + f - fn/z \end{bmatrix}. \tag{16.5}$$

Let's examine this transformation. If $z = n$ then

$$\mathbf{q} = \begin{bmatrix} x \\ y \\ n \end{bmatrix},$$

illustrating that points in the near plane do not change. If $z = f$ then

$$\mathbf{q} = \begin{bmatrix} x \cdot n/f \\ y \cdot n/f \\ f \end{bmatrix},$$

and therefore points in the far plane stay in the far plane; however, the x and y values do change. Just how they change is illustrated in Figure 16.7. We join the point in the frustum's far plane to the eye and record the point of intersection $\mathbf{q}'$ in the near plane (view plane). Then we construct a line perpendicular to the near plane at

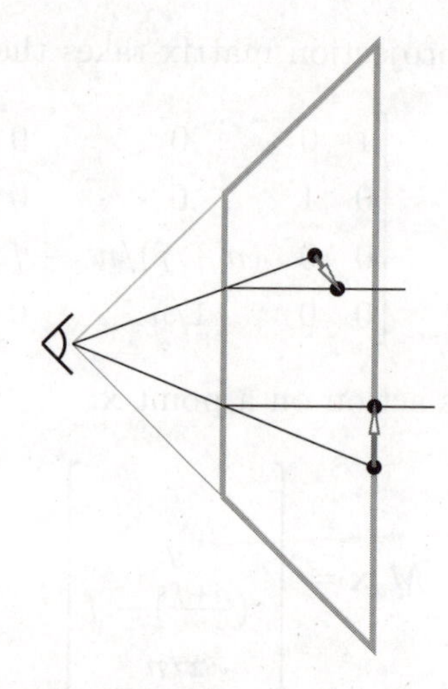

Figure 16.7. The mapping M_p "squishes" the frustum into the box as a projective map. A point is mapped to a line, which is the parallel projector through the point's image in the view plane. Except for at the near and far plane, points move away from the eye.

$\mathbf{q}'$, and intersect this line with the far plane. This is $\mathbf{q}$'s position in the box.

Further examination of this mapping reveals that points farther from the eye are scaled more in x and y. The change in z is illustrated in Figure 16.7, and this follows in the same manner as outlined above for points in the far plane.

Summarizing, points in eye coordinates are mapped to points in clip coordinates as follows:

$$\bar{\mathbf{p}}_c = M_o M_p \bar{\mathbf{p}}_e.$$

16.2.3 Perspective Division

Clipping is the process of removing all vertices (primatives) outside of the viewing volume. The vertices that remain are then projected into 3D by dividing by the homogeneous coordinate; thus,

$$\begin{bmatrix} x \\ y \\ z \\ w \end{bmatrix} \rightarrow \begin{bmatrix} x/w \\ y/w \\ z/w \end{bmatrix}.$$

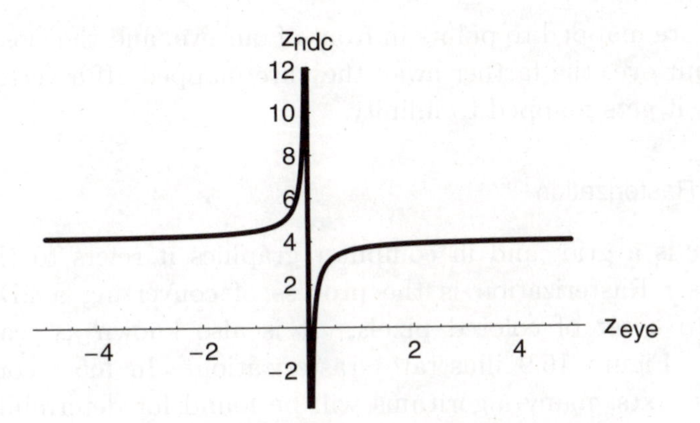

Figure 16.8. A graph of z_{ndc} as a function of z_e with $n = 0.1$ and $f = 4$. The change in z as a result of perspective division is most dramatic behind the near distance.

This is called perspective division, and the points are now said to reside in NDC coordinates.

One might ask why clipping takes place in 4D, since working in affine spaces is much more intuitive. An interesting insight into the reasoning behind clipping in 4D is presented by P. Shirley [16].

To simplify the discussion, assume that we are dealing with positive near and far distances and z-values. In (16.4) we observed that a z-value, z_e, in eye coordinates is mapped by M_p to a (homogeneous) z-value, z_c, in clip coordinates:

$$z_c = z_e \frac{n+f}{n} - f.$$

Perspective division of (16.4) resulted in (16.5), and thus

$$z_{\text{ndc}} = n + f - fn/z_e, \tag{16.6}$$

the z-value in NDC coordinates. Figure 16.8 illustrates this function.

Up to now, we have been interested in points with $z_e \in [n, f]$. But suppose some of the geometry lies outside the viewing volume. (This isn't uncommon, as we don't always want the entire world to be displayed.) The interesting aspect of Figure 16.8 is the mapping of points with z_e outside of the frustum (in z). Some points between our eye and the near plane are mapped behind our eye. Points behind our eye are mapped to points in front of our eye, and the closer they are to our eye, the farther away they are mapped. If a vertex is at our eye, it gets mapped to infinity.

16.2.4 Rasterization

A *raster* is a grid, and in computer graphics it refers to the grid of pixels. Rasterization is the process of converting a 2D primitive into a set of colored pixels. It is also known as scan conversion. Figure 16.9 illustrates rasterization. In most computer graphics texts, many algorithms will be found for determining the best selection of pixels with which to approximate a given primitive. Special methods have been developed to deal with *aliasing*, which is the appearance of stair-steps edges. These are called *anti-aliasing methods.*

Recall from Section 16.2.3 that after perspective division, we are left with 3D primitives that simply need to be projected into the xy-plane to be ready for rasterization. Thus at the rasterization step, the z-values are available, and they are stored in memory called the *z-buffer*, or *depth buffer*. Buffers are temporary storage areas for data waiting to be directed toward a device. The z-buffer and the color buffers match the raster partition. Section 16.2.5 looks at how the z-buffer is used to ensure that primitives visible from the current camera position are displayed.

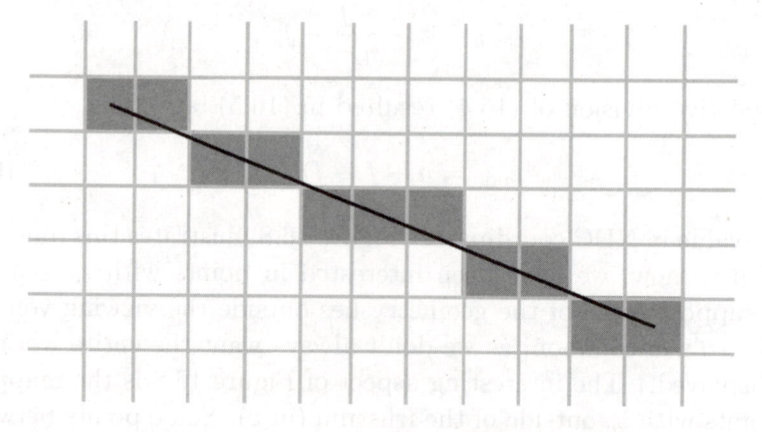

Figure 16.9. Rasterization: converting a 2D primitive into pixels.

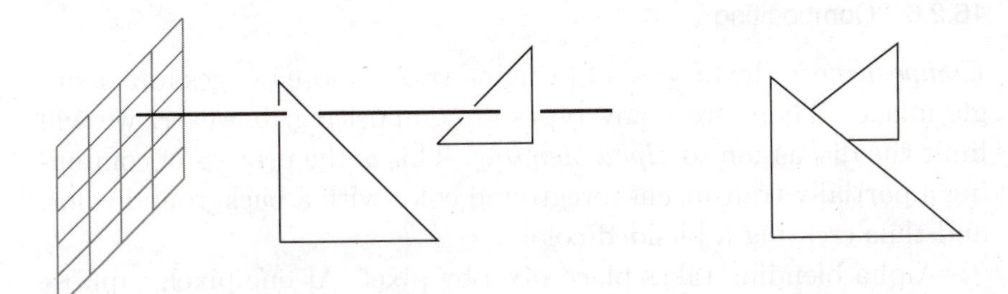

Figure 16.10. A z-buffer hidden surface removal: rasterized primitives written to a pixel are compared (left), and the rasterized primitive closer to eye is kept (right).

16.2.5 Hidden Surface Removal

Hidden surface removal is the process of determining which primitives are not visible from a given viewpoint and then removing them from the display pipeline. Figure 16.10 illustrates this process. A different perspective on this goal is called *visible surface determination.*

In the context of the graphics pipeline described in the preceding sections, the most important hidden surface removal method is the *z-buffer method.* The primitives are sent down the pipeline one at a time, but they are not sorted in any sense. As mentioned in Section 16.2.4, rasterization saves the z-value from the originating primitive in the z-buffer. Suppose a point on a primitive is currently stored at pixel p_{ij} with z-value z_a. Suppose further that the next point that is to be written to pixel p_{ij} has z-value z_b. If z_b is closer to the eye than z_a, it will replace the existing stored value.

Several hidden surface removal sorting algorithms do not depend on rasterization. *Painter's algorithm,* for example, sorts primitives from back to front. The name comes from the fact that we will draw the primitives in this back-to-front order, working similarly to a painter.

Ray tracing, which is discussed in Section 16.3.2, achieves hidden surface removal by the nature of the algorithm.

16.2.6 Compositing

Compositing is the process of bringing two or more images into a single image. There are many types of compositing; however, we will limit the discussion to *alpha blending*. This is the process of combining a partially translucent foreground color with a background color, and thus creating a blended color.

Alpha blending takes place pixel by pixel. At one pixel, suppose the background color is $\mathbf{c}_b$ and the foreground (closer to the eye) color is $\mathbf{c}_f$. If all primitives are opaque, then the pixel is colored with $\mathbf{c}_f$. However, if some primitives are partially transparent, then alpha blending is used. The new color for the pixel will be

$$\mathbf{c} = \alpha\mathbf{c}_f + (1 - \alpha)\mathbf{c}_b, \tag{16.7}$$

where α is the degree of transparency belonging to $\mathbf{c}_f$. Clearly $\alpha = 1$ corresponds to opaque and $\alpha = 0$ corresponds to completely translucent. This is the basic idea behind alpha blending; however, the blending process can take many other forms.

16.3 Illumination Models

The process of creating a 2D image from 3D geometry is called *rendering*. The steps defined in Section 16.2 are an important element of rendering, as are illumination models. It is common to interchange the terms illumination and rendering even though their exact definitions are different.

Illumination models are derived from laws of physics. For example, the so-called *rendering equation* is an integral equation that describes the flow of light energy through a scene based on the law of conservation of energy. This equation is in general too complicated to compute, and the details of the equation are not important here. However, the important idea is that rendering methods are attempts at approximating this ideal. When evaluating a rendering method, we must consider how well it models reflection, absorption, refraction, emission, and transmission of light.

In the sections to follow, we look at three rendering paradigms: real-time rendering with the local Phong illumination model, global illumination with ray tracing, and nonphotorealistic rendering.

Because of advances in processors, the boundaries of "real time" are moving. Even though real-time methods do not produce the same quality images as global methods, real-time rendering is effective for design before using valuable resources on high-end rendering methods.

16.3.1 Local Illumination: Lighting and Shading

In this section, we will look at an example of a *local illumination*, which involves a single interaction of light and objects. In contrast, global illumination, such as ray tracing, involves multiple interactions of light and objects. Because of its simplicity, local illumination can be computed in real time. OpenGL, for example, is an implementation of local illumination.

Lighting refers to a method to provide artificial illumination. *Shading* is the process of producing gradations of light or color. So first we discuss a method to produce color at the vertices of triangles based on a lighting model, and then we discuss how to shade the interior of the triangle by using the colors at the vertices. It is common to find these terms interchanged in some writings.

In this section, we assume that we are lighting and shading a 3D triangle mesh. We will need unit normal vectors at the vertices. If we have tessellated a known function, then normals are easy to obtain. If an underlying function is not given, then these normals must be approximated. In Section 13.5, a method was given for this purpose.

Phong Lighting Model. All lighting models are simply approximations of real-world lighting. Global models tend to be better heuristics than local models, and the price is computation time. The *Phong lighting model* is the most common heuristic for local lighting models.

In the following, we examine three important elements of a lighting model.

1. *Light sources.* Number, type (desk lamp or sun), color;

2. *Material properties.* Reflection and absorption of light;

3. *Reflections.* The physics of reflection.

These are the elements that give us the impression that an object on the screen is really 3D; we perceive depth through variations in color.

In order to illuminate objects, we must first invent light sources. The Phong model breaks the light's properties into three elements.

- *Ambient light.* This is scattered light with no detectable direction. Perception of this light is not dependent on the viewpoint.
- *Diffuse light.* This is directional light that scatters equally in all directions upon hitting an object. This light is not dependent on the viewpoint.
- *Specular light.* This light comes from a detectable direction, and it bounces off an object in a particular direction. It plays a role in shininess, and it is dependent on the viewpoint.

In addition, we must distinguish between light sources.

- *Point source or spotlight.* A point source emits light in all directions; a spotlight emits light in the shape of a cone.
- *Positional or directional.* A positional light source acts as a desk lamp; a directional light source acts like the sun, whereby all rays are parallel when they reach an object.

It is standard practice to compute a color with the RGB color model because that is what is used by the display hardware. Therefore, we represent the ambient, diffuse, and specular light colors by $\mathbf{l}_a, \mathbf{l}_d, \mathbf{l}_s$, respectively, where the components of each represent amounts of red, green, and blue light. Each color component can take values between 0 (no intensity) and 1 (full intensity).

Next, we must give the 3D mesh *material properties.* This determines how the material reflects and interacts with light.

- *Ambient reflectance.* Specifies the amount of ambient light reflected at a point on the object.
- *Diffuse reflectance.* Specifies the degree of scattering of light at a point on the object. A matte or gloss paint finish on the object will be the result of low or high diffuse reflectance settings, respectively.

- *Specular reflectance.* Specifies the degree of mirror-like quality
- *Transparency.* Specifies the degree of being transparent or clear. This is also known as alpha value, and written as α or A. An opaque surface will have $\alpha = 1$, and a translucent surface will have $\alpha = 0$.

The ambient, diffuse, and specular reflectance of the material are represented by $\mathbf{k}_a, \mathbf{k}_d, \mathbf{k}_s$, respectively. A component of a material property vector represents the percentage of reflection of the light source's corresponding property, and therefore these values assume values between 0 and 1.

Absorption and reflectance of light influence the perceived color of an object. For instance, if a red box is illuminated with a white light,[4] it will appear red because the red light component is reflected and green and blue light components are absorbed. However, if the red box is illuminated by a green light, it will appear black because there is no red component to be reflected.

For instance, to simulate brass we would assign the material properties as

$$\mathbf{k}_a = \begin{bmatrix} 0.32 \\ 0.22 \\ 0.02 \end{bmatrix}, \quad \mathbf{k}_d = \begin{bmatrix} 0.78 \\ 0.56 \\ 0.94 \end{bmatrix}, \quad \mathbf{k}_s = \begin{bmatrix} 0.99 \\ 0.94 \\ 0.80 \end{bmatrix}.$$

We can observe that reddish-brown light will be reflected. There is a fairly high amount of specular light reflected with a reddish tint, and therefore the surface will be shiny.

The geometric elements of the lighting model are illustrated in Figure 16.11. We are computing the color at a vertex $\mathbf{p}$. The light vector $\mathbf{l}$ is defined as the difference between the light location and $\mathbf{p}$. The normal $\mathbf{n}$ at the vertex must be computed. The reflection vector $\mathbf{r}$ is easily computed using the fact that the angle of incidence, θ, that is the angle between $\mathbf{l}$ and $\mathbf{n}$, equals the angle of reflection, or the angle between $\mathbf{r}$ and $\mathbf{n}$. The view vector, $\mathbf{v}$, is computed as the difference between the eye location and $\mathbf{p}$. The angle ϕ is the angle between $\mathbf{v}$ and $\mathbf{r}$. All vectors should be normalized.

[4]White light is composed of red, green, and blue light components (see Section 16.1).

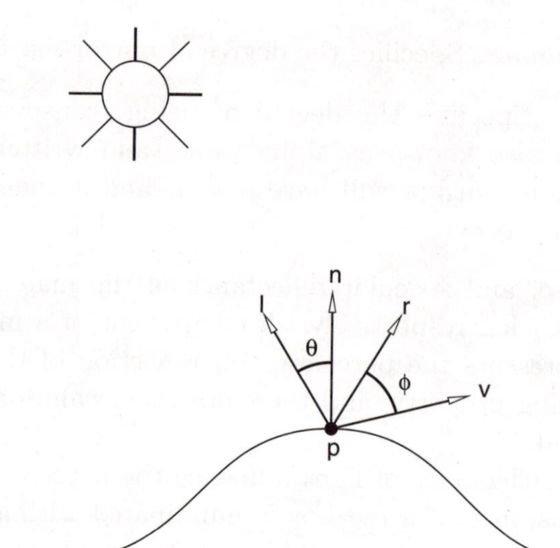

Figure 16.11. Geometric elements for lighting calculation with the Phong model.

Diffuse light intensity $\mathbf{I}_d$ at $\mathbf{v}$ is computed by using *Lambert's law*, which states that the light reflected is proportional to the cosine of the angle θ between $\mathbf{l}$ and $\mathbf{n}$,

$$\mathbf{I}_d = \mathbf{k}_d \star \mathbf{l}_d \cos(\theta).$$

Note that this equation commits an abuse of notation! A dot product is not to be taken; rather, each component (red, green, blue) should be computed independently.[5] If $\theta = 0$, then the cosine function is 1, and this results in maximal diffusion (scattering) of light. As θ approaches 90°, the amount of scattering reduces. Notice that the diffuse light intensity is independent of the viewer's location.

If we consider only diffuse light, a vertex whose normal points away from the light will be black. Ambient light allows us to illuminate the entire model. *Ambient light intensity*, $\mathbf{I}_a$ at $\mathbf{v}$ is computed as

$$\mathbf{I}_a = \mathbf{k}_a \star \mathbf{l}_a.$$

Again, the material property attenuates the light's ambient property.

[5]We will use the $\star$ notation to indicate that multiplication occurs component-wise.

Specular light intensity $\mathbf{I}_s$ at $\mathbf{v}$ is computed based on the angle between the viewpoint and reflection vector,

$$\mathbf{I}_s = \mathbf{k}_s \star \mathbf{l}_s \cos^n(\phi).$$

The power n adds control over the focus of the specular effect, or radius of specular intensity. For $n = 1$, the specular effect is fairly wide. For $n = 100$, the specular effect is rather focused, creating a sharp specular highlight. Similar to the diffuse component, the closer the viewpoint and reflectance vector become, the more accentuated the specular effect becomes.

If the object (vertex) emits light, we add material emission, $\mathbf{e}$. It is also possible to add to the scene global ambient light, $\mathbf{M}_a$, which is independent of the light sources. Thus, the global ambient intensity for an object is

$$\mathbf{I}_g = \mathbf{k}_a \star \mathbf{M}_a$$

Putting it all together, the intensity, $\mathbf{I}$, at a vertex is

$$\mathbf{I} = \mathbf{e} + \mathbf{I}_g + \sum_{\text{light sources}} (\mathbf{I}_a + \mathbf{I}_d + \mathbf{I}_s).$$

Attenuation of light based on the distance of the light source from a vertex or proximity to a spotlight is easy to add.

Shading. The Phong illumination model provides a method to calculate the color at a vertex. *Shading* methods provide the means to fill in a triangle. When the triangle is rasterized, colors must be assigned inside the triangle. We'll look at two methods that are frequently used: flat shading and smooth shading. Figure 16.12 illustrates an example of each. The top right figure displays the triangle edges; this display mode is called *wireframe*.

Flat shading is very simple: the color from one vertex is used for the entire triangle. Flat shading creates a very faceted look, but for scientific data this might be appropriate. Smooth shading, described next, creates an artificial smoothness that might mask phenomena to be observed.

Smooth shading, as the name indicates, creates a smoother looking surface because this method uses the colors computed at each vertex to interpolate color across the triangle. Specifically, suppose

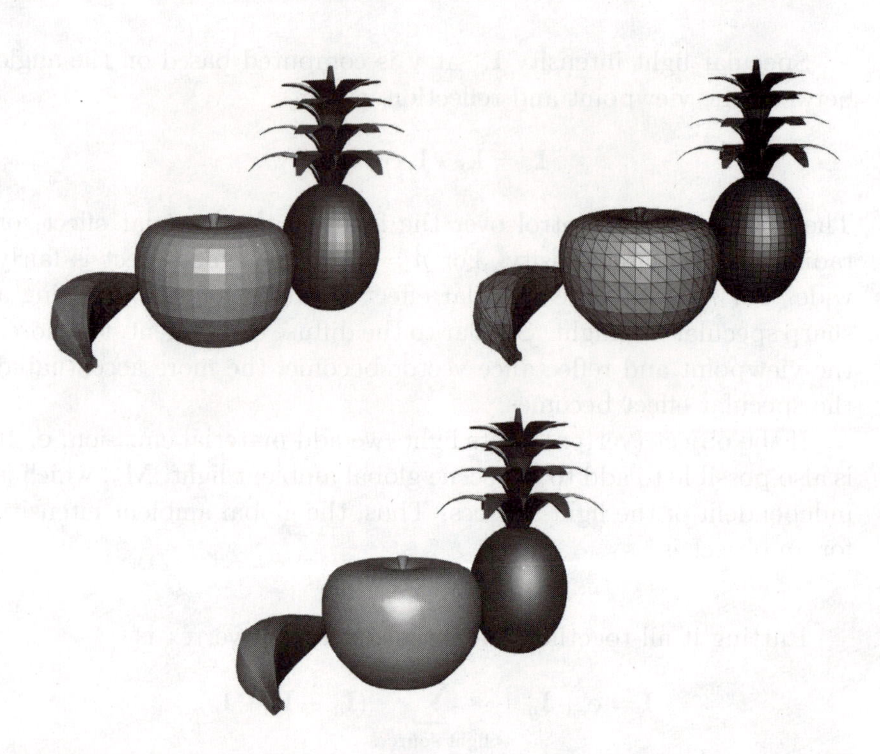

Figure 16.12. Constrast among flat shading (top left), flat shading with wire-frame display (top right), and smooth shading (bottom). (Courtesy of Kerstin Müller and Christoph Fünfzig.)

the intensity (color) at the three vertices of a triangle have been computed to be $\mathbf{I}_1$, $\mathbf{I}_2$, and $\mathbf{I}_3$. At point $\mathbf{p}$, which has barycentric coordinates (see Section 13.3.2) u, v, w with respect to the triangle vertices, the color is computed as

$$\mathbf{I} = u\mathbf{I}_1 + v\mathbf{I}_2 + w\mathbf{I}_3.$$

Smooth shading is also known as *Gouraud shading.*

16.3.2 Global Illumination: Ray Tracing

The Phong illumination model works well for creating real-time images, but it has its limitations because of the single interaction basis and its lack of global illumination considerations. A straightforward global illumination method is called *ray tracing.* In general, ray

Figure 16.13. A ray-traced scene with multiple levels of reflection and shadows.

tracing is not considered to be real time; however, some special implementation can achieve this. As illustrated in Figure 16.13, ray tracing produces reflections and shadows.

Ray tracing is based on the idea that an image is formed by the light entering each pixel. The color of this light is created by light sources interacting with objects. Computing all interactions of light in a scene would be too complex, so ray tracing works backward by casting a ray from an imaginary eye through a pixel and computing interactions. This idea is illustrated in Figure 16.14. When the

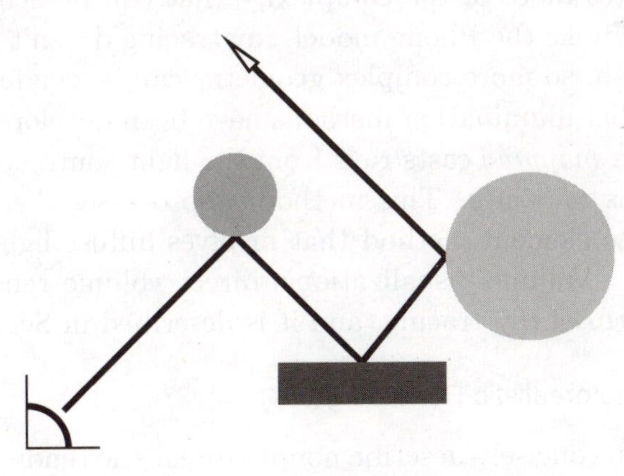

Figure 16.14. Ray tracing: a ray from the eye, through a pixel, is cast into the scene. At each intersection, lighting is computed, and the total light incident on the eye is recorded at the current pixel.

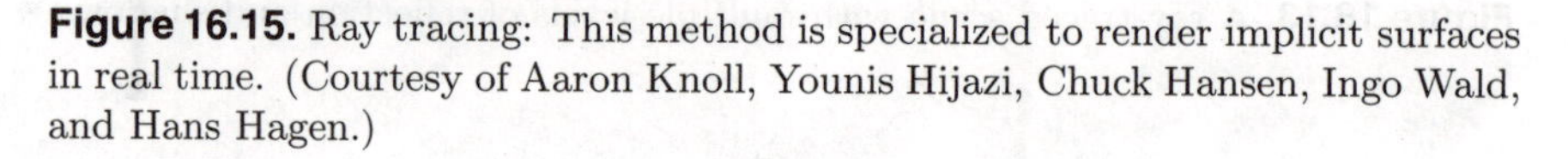

Figure 16.15. Ray tracing: This method is specialized to render implicit surfaces in real time. (Courtesy of Aaron Knoll, Younis Hijazi, Chuck Hansen, Ingo Wald, and Hans Hagen.)

cast ray intersects an object, the color at this point is computed by using a model, such as the Phong model. Depending on the material properties at this point, the ray is either reflected, absorbed, or refracted. Taking each of these possible behaviors into account, the ray continues to move through the scene. The number of intersections (recursions) computed is an input parameter to the ray tracer. Allowing only one intersection is called ray casting. Figure 16.15 illustrates more of the complexity that can be achieved with ray tracing. Unlike the Phong model, ray tracing doesn't necessitate a triangle mesh, so more complex geometry can be rendered.

Other global illumination methods have been developed. For example, *photon mapping* casts rays from the light sources. A favorite of architects is *radiosity.* This method preprocesses the scene with a type of finite element method that involves diffuse lighting calculations only. Volume visualization's direct volume rendering is a volumetric form of ray tracing, and it is described in Section 15.4.

16.3.3 Nonphotorealistic Rendering

It is difficult to concisely describe nonphotorealistic rendering (NPR) because, just as with art, there is no one definition. Much of computer graphics is consumed with reproducing the world as we see it. NPR takes an artistic approach, as illustrated in Figure 16.16.

Figure 16.16. Nonphotorealistic rendering of an engineering part. (Courtesy of Amy Gooch and Bruce Gooch.)

Our interpretation of shape comes from outlines, silhouettes, creases, and shading, and those are the primary tools of NPR. Outside of its artistic value, NPR is useful, for example, in engineering manuals, where unnecessary detail can be removed and relevant parts can be rendered to highlight them. The tools of NPR mimic those of the artworld: hatching, shading, emphasis on silhouettes. More information can be found in a text by Gooch and Gooch [6].

16.4 Texture Mapping

Texture mapping is a method to create complexity in a model without the overhead of building a large geometric model. This is done by modifying the color of the model according to the color in a given image, called the *texture*. A classic example of 2D texture mapping is the creation of a brick wall by applying a brick wall image to a single rectangular block. Figure 16.17 illustrates the result of texture mapping. Thousands of triangles would be needed to accurately model the brick wall, wood roof, and grass.

In *2D texture mapping*, we are given a triangle mesh and a texture that is defined as an $m \times n$ array of *texels*,[6] and each texel has one color assigned to it. The texture is defined in a 2D (parameter)

[6]The term texel referes to "texture element."

Figure 16.17. 2D texture maps applied to the roof, walls, and ground of a low polygon count scene (base geometry, left), result in a detailed scene (right). (Courtesy of Kerstin Müller and Christoph Fünfzig.))

space (u, v), where $a \leq u \leq b$ and $c \leq v \leq d$. A color is defined for each (u, v). Each vertex in the triangle mesh is assigned a (u, v) parameter value, or texture coordinate. There are many methods for determining how to fill the triangles given the texture coordinates at the vertices. Textures are frequently square and in sizes that are powers of two to accommodate downsizing and filtering, which are applied to produce the best texture based on the final image size. Just how difficult it is to find texture coordinates depends on the triangle mesh and the complexity of texture application desired.

Several methods are available for determining how the texture interacts with the material properties of the triangle mesh. As an example, the texture can be a decal and entirely replace the color of the object.

Color maps (see Section 14.2) are one type of texture mapping technique. This generalization of the technique uses a procedural definition of the texture, producing textures without a parametrization of the mesh.

Texture mapping can be constructed from 1D or 3D mappings as well. *1D texture mapping* is useful for contouring or cartoon (cel) shading. Cartoon shading, illustrated in Figure 16.18, is created by modifying the color at a vertex based on the angle between the normal and light source. *3D texture mapping* creates *voxel texture elements.* This type of texturing is really quite simple because one

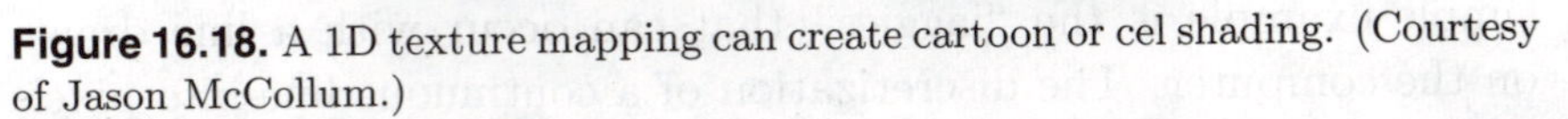

Figure 16.18. A 1D texture mapping can create cartoon or cel shading. (Courtesy of Jason McCollum.)

must simply embed the geometry in the texture space to determine the correspondence to the texture. This method is commonly used in medical and geoscience applications.

Texture mapping has its drawbacks though. For example, the brick wall in Figure 16.17 will not have a correct silhouette. Finding a good mapping (or atlas) can be challenging. In addition, the applied texture will not interact with the light sources correctly.

Other mapping techniques include *displacement maps*, in which the mapping modifies the position of the vertices, and *bump maps*, in which the mapping modifies the normals.

16.5 Sampling, Smoothing, Reduction

The goal of this section is to introduce terms and ideas that might be encountered with respect to pre- and postprocessing of data. Mathematical details on the topics in this section are beyond the scope of this text; however, an introduction to some terms in these areas is appropriate.

Data acquisition technologies *sample*. For example, a laser scanner records points and a CT scanner records average densities in voxels. In this book, we have discussed modeling and rendering of sampled data. The rendering process, by means of projecting the image to pixels, is yet another sampling process. Our eyes take samples as well!

Resampling occurs frequently in scientific computing and visualization. As we see with marching squares or marching cubes, often it is convenient to have data on a regular structure. If a data acquisition tool does not return data in a regular form, then it is resampled. With CT or MRI, data preparation can involve resampling to obtain a regular grid, to correct for patient motion, to enhance contrast, or to create additional slices. Aircraft LiDAR data provide an example of this as well. Low-flying aircraft records high-resolution DEMs by resampling measured data at regularly spaced intervals.

Aliasing is a distortion that can occur with sampling and reconstruction. The reconstruction is an alias of the original data. A simple example is the "jaggies" that can occur with a line drawn on the computer. The discretization of a continuous line into pixels can result in a stair-step appearance. Aliasing is also experienced by looking through a screen door. In this case, the artistic patterns that appear are called *Moiré patterns.* Antialiasing methods are designed to diminish the appearance of artifacts.

Great amounts of literature in signal processing, image processing, and computer graphics have been devoted to sampling theory for the purpose of determining optimal sampling rates, optimal reconstruction methods, and means to control aliasing. In 1949, Shannon proved a theorem that is now commonly known as the *Shannon sampling theorem.* This theorem states that a signal that is restricted to a particular frequency range, called *band-limited,*[7] and is sampled with a frequency at least twice its highest frequency is completely determined by its samples. This frequency is called the *Nyquist frequency.* In theory, the signal can be reconstructed by convolution (a type of moving average) with the sinc function ($\sin(\pi x)/\pi x$; the name is short for sine cardinal). The sinc function reconstruction cannot be used in practice, so many researchers work on practical reconstruction functions, or in other words, interpolating between sampling points.

One negative aspect of marching cubes is the enormous number of triangles produced and the long sliver-like shape that some take. The former problem increases rendering time and increases file size,

[7]For example, a CD player is band-limited to 20 kHz by the reconstruction filters built into the player.

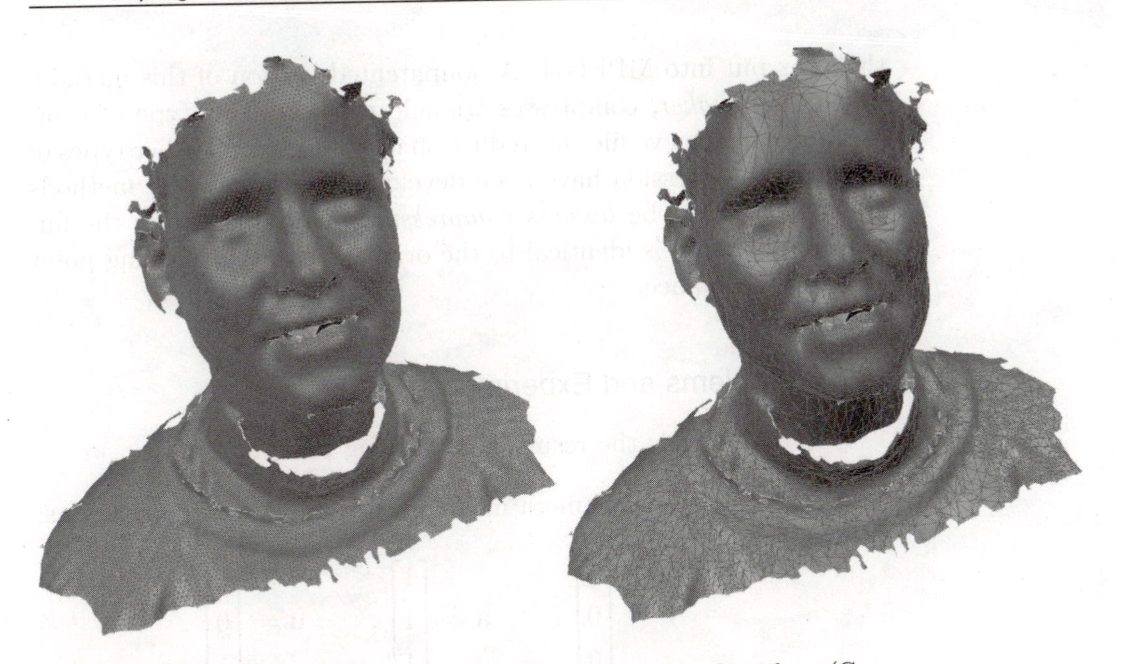

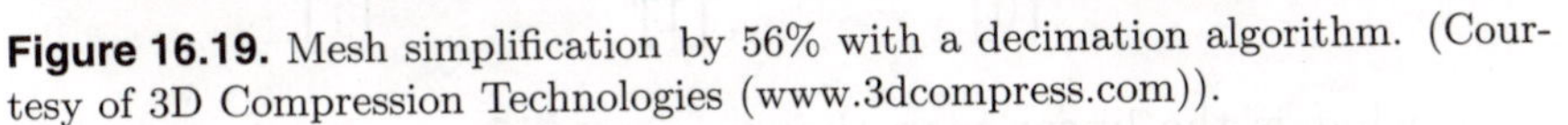

Figure 16.19. Mesh simplification by 56% with a decimation algorithm. (Courtesy of 3D Compression Technologies (www.3dcompress.com)).

resulting in longer loading times. The latter problem can cause rendering irregularities. The number of triangles produced has more to do with the volume partition and less to do with the underlying function's complexity. *Mesh simplification* is a means to reduce the number of triangles. Several strategies are possible; Figure 16.19 illustrates what is called *decimation*, in which edges are collapsed, resulting in fewer vertices and triangles. Mesh simplification methods examine local behavior of the mesh to determine areas that can be reduced without much loss in shape information. Additionally, sliver triangles are candidates for removal. Several mesh simplification methods take a different approach; they resample the mesh. Mesh simplification is a form of *lossy compression*, which is compression with information loss.

Sharing large meshes over networks can be problematic. Standard compression techniques, such as JPEG, work well for text or 2D images because they are designed to recognize repeating patterns of bytes. They are not designed, however, to optimally compress 3D geometry. In the mid-1990s, IBM patented a 3D compression scheme that was put into MPEG-4. A nonpatented version of this method, called *edgebreaker*, compresses triangle meshes. This type of compression can achieve file size reduction of 95%. Several other types of 3D mesh compression have been developed as well. These methods are considered to be *lossless compression* methods because the uncompressed mesh is identical to the original within a (floating point accuracy) tolerance.

16.6 Problems and Experiments

1. What color is the result of combining red and cyan?

2. If we set up our camera model with the following parameters:

$$\mathbf{e} = \begin{bmatrix} 0 \\ 0 \\ 0 \end{bmatrix}, \quad \mathbf{a} = \begin{bmatrix} 1 \\ 1 \\ 1 \end{bmatrix}, \quad \mathbf{u} = \begin{bmatrix} 0 \\ 0 \\ 1 \end{bmatrix},$$

   what is the linear map that maps world coordinate points to eye coordinates?

3. Experiment with the viewing, transformation, lighting, and projection tutorials on Nate Robin's OpenGL resources pages: http://www.xmission.com/~nate/tutors.html.

4. Give the 4×4 matrix M that scales by 2 in y, scales by 3 in z, and translates by $[4, 5, 6]^{\mathrm{T}}$.

5. If a green light illuminates a red box, what color does the box appear?

6. What type of light (ambient, diffuse, or specular) is independent of the viewer's position?

7. Sketch the cosine function between $-180°$ and $180°$.

8. Contrast and compare local and global illumination methods.

9. Defining a texture map over a sphere can be tricky. See a real-world example of this by experimenting with Google Earth and

exploring the polar caps. Google Earth also provides a good example of the challenge of stitching together texture maps. Identify problem areas and mixed resolution areas.

10. Multiresolution methods consider the screen space an object occupies to determine the detail of the model needed. How might mesh simplification be used to support a multiresolution method?

Bibliography

[1] G. Farin and D. Hansford. *The Essentials of CAGD*. Wellesley, MA: A K Peters, Ltd., 2000. (http://www.farinhansford.com/books/essbook.)

[2] G. Farin and D. Hansford. *Practical Linear Algebra, A Geometry Toolbox*. Wellesley, MA: A K Peters Ltd., 2005. (http://www.farinhansford.com/books/pla.)

[3] J. Foley and A. Van Dam. *Fundamentals of Interactive Computer Graphics*. Reading, MA: Addison-Wesley, 1982.

[4] George E. Forsythe, Michael A. Malcolm, and Cleve B. Moler. *Computer Methods for Mathematical Computations*. Englewood Cliffs, NJ: Prentice Hall, 1977.

[5] Karen A. Frenkel. "Volume Rendering." *Communications of the ACM* 32:4 (1989), 426–435.

[6] Bruce Gooch and Amy Gooch. *Non-photorealistic Rendering*. Wellesley, MA: A K Peters, Ltd., 2001.

[7] Michael T. Heath. *Scientific Computing: An Introductory Survey*, Second edition. New York, NY: McGraw-Hill Higher Education, 2002.

[8] Adobe Systems Inc. *PostScript Language Reference Manual*. Reading, MA: Addison-Wesley, 1985.

[9] Adobe Systems Inc. *PostScript Language Tutorial and Cookbook*. Reading, MA: Addison-Wesley, 1985.

[10] L. Johnson and R. Riess. *Numerical Analysis*, Second edition. Reading, MA: Addison-Wesley, 1982.

[11] David Kahaner, Cleve B. Moler, and Steven Nash. *Numerical Methods and Software.* Englewood Cliffs, NJ: PrenticeHall, 1989.

[12] Rubin H. Landau. *A First Course in Scientific Computing: Symbolic, Graphic, and Numeric Modeling using Maple, Java, Mathematica, and Fortran90.* Princeton, NJ: Princeton University Press, 2005.

[13] David A. Lane. *Scientific Visualization*, Chapter on Scientific Visualization of Large-Scale Unsteady Fluid Flows, pp. 125–145. Washington, DC: IEEE Computer Society, 1997.

[14] H. Pottmann, A. Asperl, M. Hofer, and A. Kilian. *Architectural Geometry.* Exton, PA: Bentley Institute Press, 2007.

[15] W. Press, S. Teukolsky, W. Vetterling, and B. Flannery. *Numerical Recipes: The Art of Scientific Computing*, Third edition. Cambridge, UK: Cambridge University Press, 2007.

[16] P. Shirley, M. Ashikhmin, S. Marschner M. Gleicher, E. Reinhard, K. Sung, W. Thompson, and P. Willemsen. *Fundamentals of Computer Graphics*, Second edition. Wellesley, MA: A K Peters Ltd., 2006.

[17] D. Shreiner, M. Woo, J. Neider, and T. Davis. *OpenGL Programming Guide*, Fourth edition. Reading, MA: Addison-Wesley, 2004.

[18] E. Tufte. *The Visual Display of Quantitative Information*, Second edition. Cheshire, CT: Graphics Press, 2001.

[19] C. Ware. *Information Visualization, Perception for Design*, Second edition. San Francisco, CA: Morgan Kaufmann, 2004.

Index

$2\frac{1}{2}D$ data, 189
2D coordinate system, 13
3D archiving, 178
3D coordinate system, 15

affine map, 24, 25
algorithm, 5
aliasing, 250
alpha blending, 252
alpha-value, 255
ambient light, 254
    intensity, 256
ambient reflectance, 254
amorphous phenomena, 224
anthropology, 54
antialiasing, 250
approximation
    B-spline least squares, 104
    least squares, 98
    piecewise linear, 123
    polynomial least squares, 94
architecture, 185
area, 167
    signed, 167
arrow plot, 227
aspect ratio, 23
associative law, 40

B-spline, 99
    basis functions, 100
    interpolation, 99
    least squares approximation, 104
    properties, 100
bar chart, 157
bark beetle, 184
barycentric coordinates, 167, 171
basis, 30
basis functions, 100
Bernstein polynomial, 99
bilinear interpolation, 135
bin, 155
bisection, 128
bisector
    two points, 171
bivariate function, 132
    contour, 198
    rectangle mesh, 185
boundary conditions, 142
boundary edge, 177
boundary value problems, 115
box plot, 159
buckyball, 174
bump maps, 263
butterfly effect, 121

cam profiles, 17
cancellation, 9

Cartesian coordinate system, 13
cartoon shading, 262
case study
    3D archiving, 178
    bark beetles, 184
    computing the square root, 127
    eigenfaces, 73
    femoral head reconstruction, 54
    fluid flow, 51
    GIS, 204
    health care, 218
    Lorenz Attractor, 120
    mixing chemicals, 41
    PageRank, 61
    Scud attack, 10
    UTM coordinates, 19
    Visible Human Project, 225
    Wilkinson polynomials, 129
cell, 210
cell phone, 183
center of gravity, 179
centroid, 166
cerebrospinal fluid, 206
CFD, *see* computational fluid dynamics
chain rule, 85
chaos, 121
circumcenter, 166
circumcircle, 166
    empty, 181
Clifford algebra, 26
closed mesh, 177
CMY color model, 236
color
    24-bit, 235
    map, 194
color model, 234
    CMY, 236
    gray scales, 236
    HSL, 235
    RGB, 234
    YIQ, 236
commutative law, 40
complexity, 6
compositing, 252
    alpha blending, 252
composition
    of functions, 80
compression, 263
computational fluid dynamics, 52, 212
computational science, 5
computed tomography (CT), 190, 211, 214
    resampling, 264
condition number, 46, 69, 70, 98
conformal map, 19
connectivity matrix, 62
continuous function, 87
contour, 138, 146
    3D function, 215
    comparison with DVR, 224
    image segmentation, 206
    line, 197
    marching cubes, 215
    rectangle mesh, 185
    shaded, 204
    trilinear function, 217
convergence
    discrete Laplace solution, 143
    integrals of bivariate functions, 135
    iterative system solvers, 51
    Jacobi iteration, 64
    Newton-Raphson method, 126
    power method, 61
convex curve, 103
convex hull, 182
convex polygon, 172
convolution, 264
coordinate plane, 13
coordinate system
    barycentric, 167
    Cartesian, 13
    conversion, 19
    cylindrical, 16
    polar, 16
    spherical, 16
coordinate transformation
    distortion, 23
    local and global, 22
coordinates, 14
    angular, 16

azimuthal, 16, 17
barycentric, 167
Cartesian, 13
cylindrical, 16
Earth, 18
global, 21
latitude and longitude, 18
local, 21
normalized, 23
polar, 16
radial, 16, 17
screen, 24
spherical, 16
zenith, 17
correlation coefficient, 151
Cramer's rule, 43
crystallography, 174
CT, *see* computed tomography
cubic, 86
curvature, 180
cylindrical coordinate system, 19

data
gridded, 189
point cloud, 192
scattered, 189
data acquisition, 263
LiDAR, 190
outlier, 98
data cutting, 226
data fitting, 91
decimation, 265
Delaunay mesh, 181
DEM, *see* digital elevation model
density plot, 199
depth buffer, 250
derivative, 84, 107
Descartes, Ren/'e, 13
design matrix, 96
determinant, 39, 45, 66, 75
eigenvalues, 66
diagonal matrix, 39, 71
diagonally dominant, 51
differentiable, 84
diffuse light, 254
intensity, 256
diffuse reflectance, 254
digital elevation model (DEM), 190, 204
digitial Michelangelo project, 178
digitizer, 54
dimension, 29, 30, 87
direct volume rendering, 219, 260
comparison with contouring, 224
full volume, 222
image-order, 222
maximum intensity projection, 222
object-order, 222
ray casting, 219
sampling, 219
splatting, 222
transfer function, 222
x-ray, 222
directional light, 254
displacement maps, 263
displacement plot, 228
distance
perpendicular, 170
point and plane, 170
distortion, 23
distributive law
matrices, 40
dominant eigenvalue, 60
dominant line, 153
dot product, 37
double integral, 135
dyadic matrix, 35, 40
dynamic processes, 109
dynamical system, 118

edge detection, 206
edgebreaker, 266
eigenface, 73
eigenfrequency, 75
eigenfunction, 76
eigengene, 75
eigenvalue, 58
computation, 63
determinant, 66
dominant, 60

eigenvector, 58
  dominant, 60
  tensor ellipsoid, 230
empty circumcircle, 181
end condition, 103
equilateral triangle, 167
equilibrium, 88
error
  cancellation, 9
  discretization, 52
  truncation, 8
Euclidean norm, 46
Euler equations, 212
Euler's law, 174
Euler's method, 111
Euler-Lagrange PDE, 144

face recognition, 73
facet, 165
false color, 195
FEM, *see* finite element method
femoral head, 54
filter, 206
finite differences, 115
finite element method, 48, 213
first-order ODE, 111
flat shading, 257
floating-point number, 6
flow problems, 17
fox and rabbit problem, 118
frequency, 75, 156
full volume rendering, 222
function
  bivariate, 132
  bounded, 78
  continuous, 78, 87
  convex, 78
  definition, 77
  discrete, 78
  limit, 80
  monotone, 78
  symmetric, 78
  trivariate, 145
function space, 86

Gauss curvature, 180
Gauss elimination, 43, 172
Gauss, C. F., 43
Gauss-Jacobi iteration, 50, 51
Gauss-Seidel iteration, 51, 143
genus, 177
Geographic Information Systems, 135, 204
geometry
  projective, 25
GIS, *see* Geographic Information Systems
global illumination
  direct volume rendering, 260
  photon mapping, 260
  radiosity, 260
  ray tracing, 258
glyph, 228, 230
Google, 62
Gouroud shading, 257
gradient, 133
graph
  1D, 78
  2D, 131
  3D, 145
graphics pipeline, 236
gravitational problems, 17
grid generation, 212
gridded data, 189

half-plane, 182
haptic visualization, 230
hedgehog plot, 227
Heun's method, 113
hidden surface removal, 192, 251
  z-buffer method, 251
  painter's algorithm, 251
  ray tracing, 251
Hilbert matrix, 56, 76
histogram, 81, 155, 223
hole, 177
homogeneous coordinates, 26
HSL color model, 235
Hubble telescope, 214
hyperboloid, 140

ill-conditioned linear system, 46
illumination methods, 252

illumination model, 253
    ambient light, 254
    ambient light intensity, 256
    ambient reflectance, 254
    diffuse light, 254
    diffuse light intensity, 256
    diffuse reflectance, 254
    directional light, 254
    global illumination, 258
    material properties, 254
    nonphotorealistic rendering, 260
    Phong lighting, 253, 255
    positional light, 254
    specular light, 254
    specular light intensity, 257
    specular reflectance, 255
    spotlight, 254
image vector, 31
image-order technique, 222
imaging
    CT, 211, 214, 218, 223
    fMRI, 214
    MRI, 2, 214
    PET, 214
incenter, 166
incircle, 166
inequality
    triangle, 47
information visualization, 3
initial value problem, 110
inside/outside test, 168
integral, 105
    double, 135
interpolation, 92
    B-spline, 99
    bilinear, 135
    linear, 79
    piecewise linear, 92
    trilinear, 147
interquartile range, 160
intersection
    three planes, 171
interval, 155
inverse matrix, 36, 43
isoline, 138, 197
isosurface, 215
iterative method
    bisection, 128
    convergence, 51
    convergence rate, 51
    Euler, 111
    Gauss-Jacobi, 41, 51
    Gauss-Seidel, 51, 143
    Heun, 113
    Jacobi, 63
    Newton-Raphson, 125
    power, 60
    residual vector, 51

Jacobi iteration, 63
Jacobian, 141
jaggies, 264
JPEG, 265

knot, 93
knot sequence, 100
Kronecker delta, 38

L^1 norm, 47
L^2 norm, 46
Lambert's law, 256
Laplace's equation, 142
latitude, 18
least squares, 53, 153
    approximation, 96, 98
    error, 96
Levoy, M., 178
LiDAR, 190
    resampling, 264
light sources, 254
lighting, 253
lighting equation, 221
limit, 46, 80
linear combination, 28, 79
linear dependence, 29
linear independence, 29, 30
linear interpolation, 79
linear map
    linearity property, 33
linear ODE, 115
linear space, 27
linear system, 41
    Gauss-Jacobi iteration, 50

iterative method, 48
overdetermined, 53, 95, 96, 99, 104
sparse, 48
unstable, 46
local coordinate system
extents, 23
local illumination
Phong lighting, 253
longitude, 18
Lorenz Attractor, 120
lossless compression, 266
lossy compression, 265
Lotka-Volterra model, 119
lower quartile, 160

magnetic resonance imaging, 2, 206, 214
resampling, 264
Malthus, T., 109
Manhattan norm, 47
mapping
bump, 263
displacement, 263
texture, 261
marching cubes, 215
resampling, 264
triangle explosion, 264
marching squares, 202
resampling, 264
mask, 143, 207
material classification, 223
material occupancy, 223
material properties, 254
matrix, 27
orthonormal, 67
associative law, 40
commutative law, 40
condition number, 46
determinant, 45
diagonal, 39, 71
diagonally dominant, 51
distributive law, 40
dyadic, 35, 40
Hilbert, 56, 76
identity, 34
inverse, 36, 43
nonsquare, 67, 70
orthonormal, 38, 59
positive definite, 138
product, 34
pseudoinverse, 70
rank, 36
singular, 36, 40, 66, 70
sparse, 48, 52, 104
stochastic, 63
symmetric, 59
upper triangular, 44
Vandermonde, 94
matrix multiplication, 40
maximum intensity projection, 222
median, 160
medical imaging, 206
Mercator projection, 19
mesh, 174
3D, 186
quadrilateral, 185
rectangle, 185, 190
tetrahedral, 210
triangle, 52, 174, 190, 218, 228, 253
mesh simplification, 265
minimum, 134
minmax box, 21
Moiré patterns, 264
monomial, 87
monomial form, 129
MPEG-4, 266
multidimensional transfer functions, 224
multiple root, 130
mutually orthogonal, 38

NaN, 7
Navier-Stokes equations, 212
nearest neighbor function, 220
neighbor, 176
Newton-Raphson method, 125
nonphotorealistic rendering, 260
norm
L^p, 46
Euclidean, 46

Manhattan, 47
vector, 46
normal equations, 53, 72, 96, 99
normal vector, 166, 177, 253
at a vertex, 177
normalized coordinates, 23
numerical derivatives, 107
numerical error, 7
numerical integration, 105
Nyquist frequency, 264

object-order technique, 222
ODE, *see* ordinary differential equation
opacity, 221, 223
opaqueness, 255
operations
linear, 27
optical properties, 221
orbit
stable, 119
ordinary differential equation, 109
boundary value problems, 115
Euler's method, 111
first order, 111
Heun's method, 113
initial value problem, 110
linear, 115
predictor method, 113
predictor-corrector method, 113
orthographic projection, 225, 242, 243
marching cubes, 218
orthonormal, 38, 67
orthonormal matrix, 59
outlier, 98, 160
overdetermined system, 95, 96, 99, 104

PageRank, 61
paraboloid, 140
parameter, 22
partial derivative, 133
partial differential equation, 48, 52, 142, 175
particle tracing, 228
PCA, *see* principal components analysis, 73
PDE, *see* partial differential equation
perpendicular distance, 170
perspective projection, 225, 242, 244, 248
marching cubes, 218
PET, *see* positron emission tomography
phantom knot, 100
Phong illumination model, 253, 260
photon mapping, 260
pie chart, 158
piecewise linear approximation
roots, 123
pivoting, 44
pixel, 24, 155, 210
plane, 132
bisector of two points, 171
definition, 169
distance to a point, 170
distance to origin, 169
Euclidean definition, 171
family, 170
implicit form, 169
intersection of three planes, 171
normal, 169
parametric form, 170
point normal equation, 169
point
distance to a plane, 170
point cloud, 165, 192
polar coordinate system, 16
polygon, 172
polyhedron, 174
polynomial
Bernstein, 99
interpolanting, 92
Wilkinson's, 129
population growth, 109
positional light, 254
positive definite, 138
positron emission tomography, 206, 214
power method, 60, 63
predator-prey, 117
predictor method, 113
predictor-corrector method, 113
Prewitt filter, 206

primitive, 237
principal components analysis, 72, 153
product rule, 85
projective geometry, 25
pseudocolor, 195
pseudoinverse, 70

quadrant, 14
quadratic form, 137
quadrilateral mesh, 185
quartile, 160

radiodensities, 211
radiosity, 260
rank, 36
rasterization, 250, 251
ratio
  aspect, 23
rectangle mesh, 185, 190
rectilinear grid, 186, 209
reduction, 263
reflections, 254
region growing, 181
regression line, 152
rendering, 252, 253
  cartoon shading, 262
  global illumination, 258
  nonphotorealistic rendering, 260
  Phong lighting model, 253
resampling, 227, 264
residual vector, 51
RGB color model, 234
RGBA model, 234
right-hand rule, 13, 166
root, 123
  bisection, 128
  multiple, 130
  Newton-Raphson, 125
  piecewise linear approach, 123
  secant method, 127

saddle-shaped, 180
sampling, 263
scalar product, 37
scale invariance
  correlation coefficient, 152
scatter plot, 150
scattered data, 189
scientific visualization, 2
scientific discovery process, 1
screen coordinates, 24
Scud missile, 10
secant method, 127
segmentation, 180
sequence, 46
shaded contour plot, 204
shaded relief, 194
shaded relief map, 194
shading, 192, 253
  flat, 257
  smooth, 257
Shannon sampling theorem, 264
signed area, 167
Simpson's rule, 106
singular matrix, 36, 40, 66
singular value decomposition, 45, 64, 67
  ill-conditioned matrix, 96
slope, 110
smoothing, 263
sparse matrix, 48, 52, 104
specular light, 254
specular light intensity, 257
specular reflectance, 255
spherical coordinates, 17
splatting, 222
spline function, 99
spotlight, 254
square root, 127
stability, 46
stable orbit, 119
standard deviation, 150
star, 179
step length, 112
stochastic matrix, 63
streamlines, 228
stress tensor, 230
subspace, 31
surface extraction, 215
SVD, *see* singular value decomposition
symmetric matrix, 59

Tacoma Narrows bridge, 57
tangent plane, 133
tangent vector, 133
target box, 22
Taylor expansion, 91
tensor ellipsoids, 230
tensor field, 229
  stress, 230
tetrahedral mesh, 186, 210
texture mapping, 195, 261
  1D, 262
  2D, 261
  3D, 262
  contouring, 262
  texel, 261
Thiessen polygon, 183
time series, 149
tomograms, 219
torus, 177
transfer function, 222
  1D, 224
  multidimentional, 224
transparency, 221, 223
  alpha-value, 255
trapezoid rule, 105
triangle, 165
  area, 167
  centroid, 166
  circumcenter, 166
  equilateral, 167
  incenter, 166
  neighbor, 176
  shape, 167
triangle inequality, 47
triangle mesh, 52, 174, 228, 253
  3D archiving, 178
  analyze, 179
  closed, 177
  compression, 265
  curvature, 180
  PDE, 175
  star, 179
triangulation, 174, *see also* triangle mesh
trichromatic theory, 234
trilinear interpolation, 147, 221
trivariate function, 145
  tetrahedral mesh, 186
truncation error, 8

unit normal, 166
unit sphere, 37
unit square, 133
unit vector, 37
unstable linear system, 46
update mask, 143
upper quartile, 160
upper triangular matrix, 44
UTM coordinates, 19

Vandermonde matrix, 94
vector
  unit, 37
  zero, 28
vector field, 111, 117, 227
  displacement plot, 228
  hedgehog plot, 227
  warping, 228
vector norm, 46
  L^2, 46
vector sequence, 46
  Gauss Seidel iteration, 51
  Gauss-Jacobi iteration, 49
vertex normal, 177
viewing
  marching cubes, 218
viewport, 237, 245
Visible Human Project, 225
visible surface determination, 251
visual analytics, 3
visualization
  empirical data, 149
  haptic, 230
  scalar values over 2D data, 189
  scalar values over 3D data, 209
  scale problem, 157
  tensor fields, 229
  vector fields, 227
volume element, 209
volume rendering, 209

volume visualization, 209
  contouring, 215
  data cutting, 226
  direct volume rendering, 219
  ray casting, 219
  tensor fields, 229
  transfer function, 222
  vector fields, 227
Voronoi diagram, 182
voxel, 209

warping, 228
Wilkinson's polynomial, 129
wireframe, 257

x-ray rendering, 222

YIQ color model, 236

z-buffer method, 251
zero vector, 28